Metallography Series

Optical Microscopy of Metals

Optical Microscopy of Metals

R. C. GIFKINS BSc (Lond.), DSc, FIM

C.S.I.R.O., Melbourne, Australia

AMERICAN ELSEVIER PUBLISHING COMPANY, INC.
New York 1970

Published in the United States by
American Elsevier Publishing Company, Inc.
52 Vanderbilt Avenue
New York, New York 10017

Standard Book Number: 444-19667-6
Library of Congress Catalog Card Number: 72-109133

Made in Great Britain

Preface

It is usual in a preface to state the audience for whom the book is designed, although this can very easily mislead unless the necessary generalizations are sufficiently qualified. However, some attempt to define aims and scope must be made. This book is meant to help the practising metallographer make the best use of his microscopes by giving him an understanding of the principles on which they are based, together with a translation of these principles into operational precepts. Since many students of metallurgy are destined to become practising metallographers, it is hoped that this book will also be of use to them.

The vast majority of problems which face the metallographer can be solved (given adequate preparation of the sample) with the aid of normal bright-field microscopy, using a modest microscope. Moreover, even when that microscope has been neglected, as will often be the case, the performance will be adequate for everyday problems. However, there will be occasions when important information can be obtained only by exploiting the capability of the microscope to the full or by extending its sensitivity through some special adaptation. Fortunately the modern metallographer has a wide range of such special methods at his disposal. These will be described and their virtues and limitations given; these assessments have been made, almost without exception, on the basis of personal experience gained using the various techniques.

It must be appreciated that optical metallography is but a facet of the whole; metallography embraces a very broad range of techniques, some of which extend it by using other wavelengths than those associated with the visible spectrum, and some of which are complementary to it. This book will do no more than call attention to these and direct the reader to sources of further information. The important point is that these techniques all form part of an array of tools for use in uncovering the structure of metals, alloys and an expanding list of other opaque materials of engineering interest. Selecting the right tool for the job requires some knowledge of the construction and operation of the tool, and the intention in this book is to supply just this in relation to metallurgical microscopes.

The mathematical content of the text has been deliberately kept to a minimum, even though mathematics is often the only way of making a rigorous statement about the behaviour of light waves. Many of the formulae which have been included have been developed rather than derived; the aim here, as elsewhere in the book, has been to show how various physical factors affect the situation being considered. In this way, anyone who is prepared to take a little trouble to follow the argument can find out "why" as well as "how" with respect to the techniques of optical metallography he is using.

It is hoped that this has led to a text that can be used as a working guide to intelligent—and therefore better—metallography.

R.C.G.

Acknowledgements

Several friends and colleagues have been kind enough to read and criticize the manuscript of this book in whole or in part. This has undoubtedly led to avoidance of many errors and to greater clarity in many places. Faults which remain must be attributed to the author's obstinacy in rejecting proffered advice, despite honest attempts to see his work as others see it.

I wish especially to mention Professor M. Hatherly, whose reading of the entire manuscript was characteristically meticulous. Professor M. E. Hargreaves has given a great deal of practical encouragement, without which it would have been extremely difficult to have produced the manuscript.

Acknowledgement is also made to the following for supplying photographs for illustrations or for permission to reproduce illustrations from the sources indicated:

Dr. L. E. Samuels (Figs. 11, 12, 17, 19, 22, 39, 114); Editor "Metallurgia" (Fig. 93); Australian Iron and Steel Ltd. (Fig. 101); Carl Zeiss, Jena, East Germany (Figs, 13, 14); Bausch and Lomb, Rochester, U.S.A. (Fig. 16); Carl Zeiss, Oberkochen, West Germany (Figs. 48, 69, 117); E. Leitz, Weltzar, West Germany (Figs. 103, 110); Metals Research, Cambridge, and Jarrel-Ash, Waltham, U.S.A. (Fig. 111).

Contents

1 Introduction

The first stage in metallography is the production of a true and undistorted structure on the surface of the specimen to be examined. This usually involves mechanical or electrolytic polishing, the principles of which are sufficiently understood for reliable results to be obtained,[1,2] followed by etching, which is still very much a matter of using recipes. On the other hand, the structure will sometimes have been developed on a previously polished surface by deformation, or it may be the result of damage to a machined part; in such cases the art is more in preservation than in preparation of the specimen. However produced, the majority of specimens will require examination of a magnified image of their structure before much useful information about them can be obtained, and this generally requires the use of a microscope. It is perhaps misleading to think of the microscope merely as a magnifier, for this ignores subtleties in the formation of an image by the microscope and thereby may result, at worst, in misinterpretation or, at best, in failure to extract all the available information. With a proper appreciation of the way in which the image is formed a great deal of additional understanding can be gained about the structure examined, either by the efficient use of the basic microscope or by the use of various pieces of the auxiliary equipment which is now available.

The plan of the earlier chapters of this book is to lead up to the special techniques of metallurgical microscopy by first recalling some basic ideas on diffraction and developing these in a theory of the microscope (Chapter 2), then, in the light of this theory, describing the construction of the metallurgical microscope and the function of its components (Chapter 3), followed by a discussion of its operation to obtain optimum results (Chapter 4). In Chapter 5 the basic theory of the first two chapters is used to describe a group of techniques related to normal, bright-field illumination; these include phase-contrast. Subsequent chapters deal with techniques which rely on other basic properties of light, namely two-beam and multiple-beam interference (Chapter 6) and polarized light (Chapter 7). The theory and practice of photography applied to metallography are reviewed in Chapter 8. Chapter 9 contains a miscellany of short topics of importance, including some notes on the basic types of etching and on macrophotography, and Chapter 10 is devoted to a discussion of the elements of quantitative metallography with particular emphasis upon the measurement of grain size. There is a short concluding section

(Chapter 11) in which is presented a compendium of associated metallographic techniques.

REFERENCES

1 L. E. Samuels, *Metallographic Polishing by Mechanical Methods*, London, 1967 (Pitman).
2 W. J. McG. Tegart, *The Electrolytic and Chemical Polishing of Metals in Research and Industry*, London, 1959 (2nd edn) (Pergamon).

2 Basic Optical Theory

Not all the information carried by light reflected from a specimen can be decoded by the eyes, whether these are aided by a microscope or not. This is because our eyes are sensitive to differences in the *amplitude* of the light—which we interpret as brightness—and in its *wavelength*—which we see as colour. Reflected light often has further information locked up

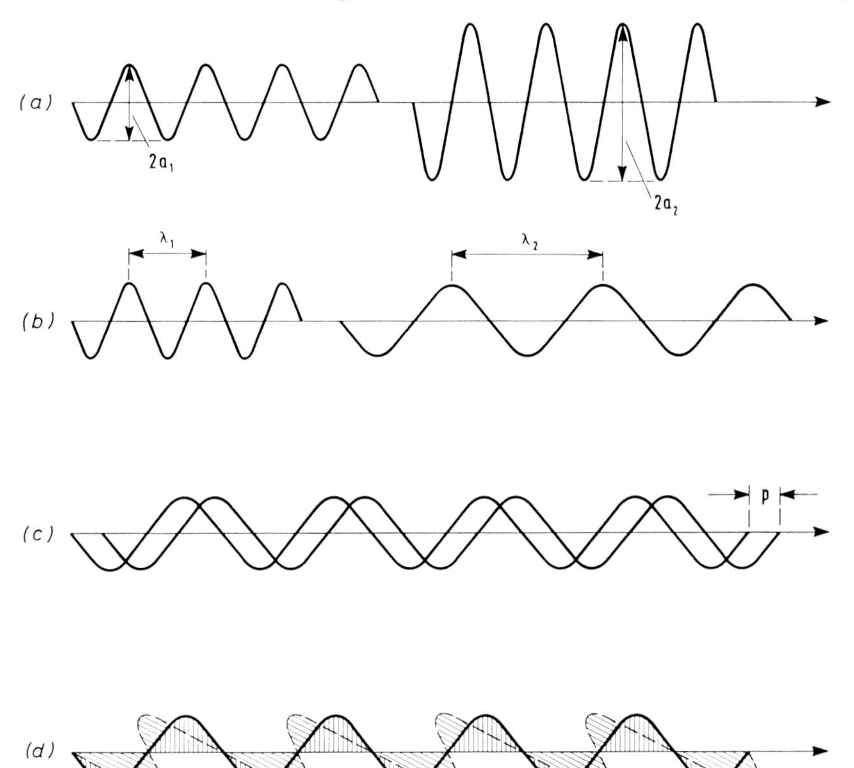

Fig. 1. The kinds of message carried by light waves (schematic).
(*a*) Amplitude: determines intensity; the greater the amplitude (a_1 or a_2) the brighter the light.
(*b*) Wavelength: determines colour; shortest wavelengths (λ_1) are blue, longest (λ_2) red.
(*c*) Phase: characterized by displacement p along the time axis; is not directly perceived by the eye.
(*d*) Polarization: describes the plane of the vibration; is not directly perceived by the eye.

in other aspects of its wavelike nature; these are the *phase* of the wave and its plane of vibration or degree of *polarization*. Some insects and animals appear to be able to distinguish between various directions of polarization, but man cannot do so (at least, not consciously) and no eyes appear to be able to detect differences in phase. Fig. 1 shows diagrammatically these four attributes of light.

Amplitude is measured by the peak height of the wave above the mean (a_1 or a_2 in Fig. 1 (*a*)), and brightness or intensity is proportional to the square of the amplitude.

Wavelength is the distance between similar points on consecutive waves, shown in Fig. 1 (*b*) from peak to peak as λ_1 or λ_2.

Phase is a much more subtle idea, which expresses the notion of the dynamic nature of a wave propagating through space. If the wave in Fig. 1 (*c*) were indeed moving forward, a particular point in space could be typified by a slot cut in a sheet of card and placed so that the wave passed across the slot. The segment of the wave seen through the slot would then appear to move up and down the slot, and the position of the segment with respect to some origin on the slot could be taken to denote the phase. In practice this idea is taken further, and the moving segment is compared with the angular position of a pointer making one revolution of a dial for every complete up-and-down motion of the segment—i.e. for the passage of one wavelength. In this way the phase now becomes characterized by the angle swept out by the pointer, and one wavelength is equivalent to a phase change of one revolution which, in turn, is equal to 2π radians or 360°.

Polarization, as shown in Fig. 1 (*d*), is connected with the orientation of the vibration with respect to the axis of propagation, but its full meaning will be deferred until Chapter 7 because some other background information on polarization effects is required before its application to metallography can be considered.

For a comprehensive treatment of these introductory ideas, one of the standard texts should be consulted, such as are listed in references 1–3.

Limits to Detail Seen

In practice we do not depend on colour for the interpretation of the vast majority of micro-specimens. Instead we depend upon variations of amplitude—in other words, on contrast—to see the details of interest. Even so, there is a further limitation, because the eye is not able to distinguish between intensities unless they differ by a certain minimum amount. Only rarely will this be a limiting factor in metallography, but there are occasions when it does become critical. Then it will be found that changing the kind of illumination or perhaps examining the negative instead of the print will reveal detail otherwise overlooked—which is the reason for the oft-heard cry of a lecturer who assures his audience that detail invisible on the screen "was clearly seen in the original negative". This is a complex problem, the visible details depending upon the kind of image considered, e.g. whether it is a white spot on a black background or vice versa. A full discussion is given by Françon.[4] Often the contrast is a result of changes in *texture*, which a higher magnification may reveal as a fairly regular repetition of detail. ("Texture" is here used in the sense it would be by an architect or decorator, and is not meant to relate to preferred crystal orientation). Unless this detail has been formed by some kind of staining, it is likely to be revealed by a shadowing effect—or what at first sight looks like shadowing—but it may turn out to be much more complex.

Let us suppose that we are looking at a specimen showing some characteristic structure when examined at a low magnification. If there is sufficient contrast we might expect to be able to isolate the basic detail of the structure by continually increasing the magnification. We would reach this conclusion by considering the formation of the magnified image to be a matter of geometry (see Fig. 2 (*a*)). In principle, if we could

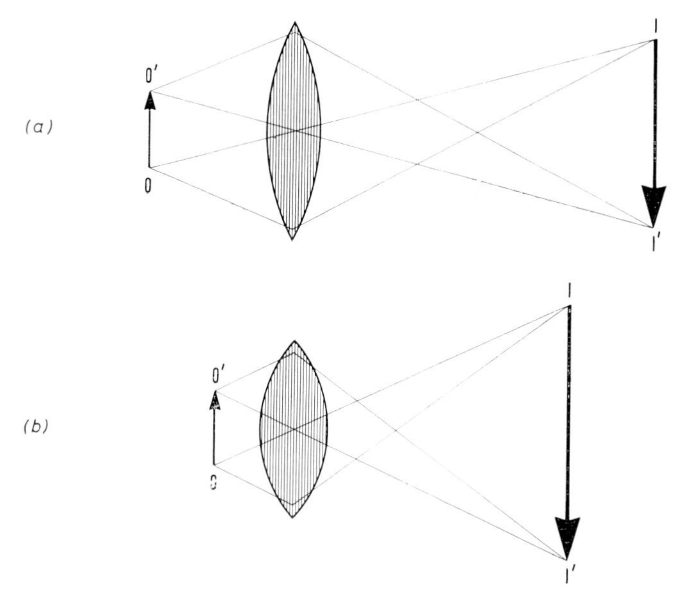

Fig. 2. Showing how the principles of geometrical optics appear to allow increasingly large magnifications (object OO′, image II′) merely by changing the shape of the lens—compare (*a*) and (*b*).

bend (i.e. refract) the rays from the object sufficiently, by changing the shape of the lens or its refractive index (Fig. 2 (*b*)), we could expect to achieve the desired *resolution* (see p. 13) of the detail. Practical experience would not bear out our expectation. There would come a stage when any further increase in magnification would not result in better definition of the detail; two closely spaced features would always remain merged. The limit of resolution for a particular microscope and wavelength of light is in fact reached when the object points are separated by a distance comparable with the wavelength of the light employed. Why this should be so requires that some basic ideas concerning diffraction of light be clearly appreciated. These will be given in the following pages in as simple a manner as possible, without sacrificing too much of the rigour which only a mathematical treatment can give. It must be emphasized at the outset that the magnification and resolution of a microscope are fundamentally different properties. They are often confused, or at least it is implied that increased magnification brings increased resolution. It will be clear in what follows that this is not necessarily so.

Diffraction and Interference

A good starting point in revealing some of the elementary ideas about diffraction of light is *Huygens' Principle*, which states that the propagation of a beam of light can be thought of in terms of secondary wavelets spreading out from imaginary elementary sources along the wavefront at a particular time. This is illustrated in Fig. 3 for a beam limited by a

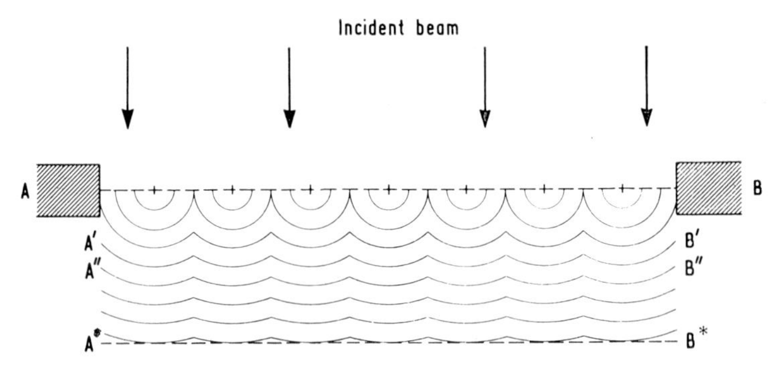

Fig. 3. Illustrating Huygens' Principle for the propagation of a beam of light. The wavefront (A′B′ . . . A*B*) at given times is constructed as the envelope of the fronts of a series of imaginary wavelets emanating from a previous wavefront AB.

wide slit and so having a plane wavefront AB, which moves to A′B′, then A″B″, etc., after successive intervals of time. The circles, centred on points on AB, represent the wavelets from the elementary sources, and the new wavefront results from their *destructive interference*. If wavelets from two neighbouring sources are considered, as in Fig. 4, it is clear that the intensity of light at a point P ahead of the previous wavefront will depend upon the relative phases of the two wavelets at that point. If the crests of the wavelets coincide, they will reinforce each other to give brightness, as in Fig. 4 (*a*); if a crest coincides with a trough, there will be cancellation and darkness, as in Fig. 4 (*b*).

On a macroscopic scale (i.e. large compared to the wavelength of light) the geometry of this situation results in the propagation of the light as a beam, just as we would expect from simple geometrical optics based on the idea of light travelling in straight lines and casting sharp shadows. On the fine scale, the edge of the beam will become ill-defined, because of the

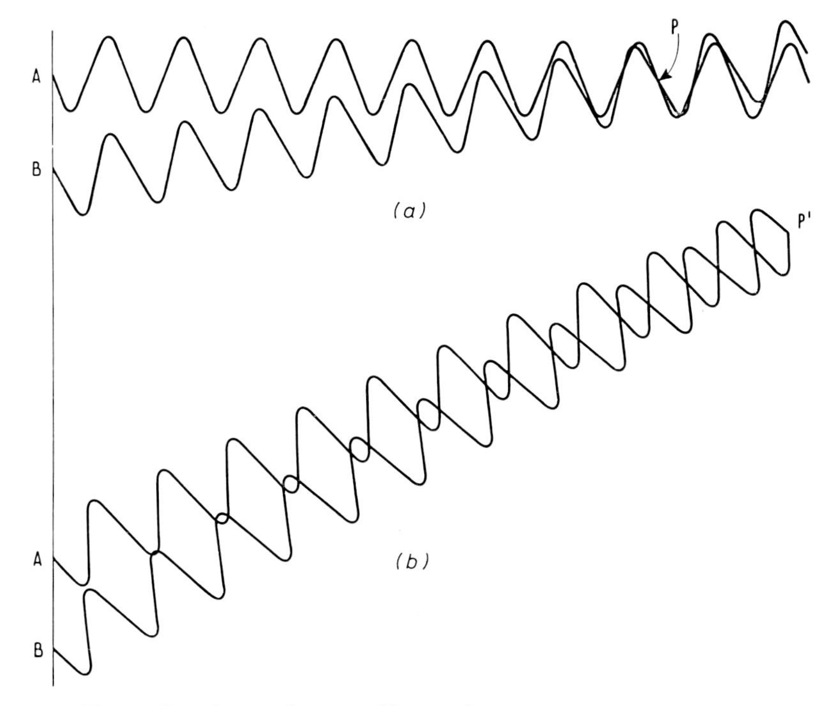

Fig. 4. Interference between Huygens' sources.
(*a*) The wavelets arrive in phase at P and reinforce one another, giving brightness.
(*b*) They are of opposite phase at P′ and so cancel, giving darkness.

diffraction or the "escape" of light by the sideways emanation from the secondary sources. It is this kind of fine-scale effect which will be seen to be important in microscopy.

Coherency

The interference effects just described are obtained only when the light rays involved are coherent. This term describes the vibratory conditions of the rays and, in its strictest sense, means that the vibrations of all the rays are in phase. In general, this can be achieved only when the rays originate from precisely the same emitter of light—that is, one atom. However, emitters (atoms) may be constrained to vibrate together in phase in a device called a laser, and so very powerful coherent sources may be created.

The true coherence which is obtained with laser light, as well as the enormously high energy densities, leads to many interesting properties of such beams. These have already been exploited in a metallurgical context to reveal very small changes in surface topography caused by strain. It is very likely that lasers will have further uses in metallography, either as tools for causing intensive local heating of micro-specimens or as a means of obtaining three-dimensional photography by the techniques of holography.

The phases of all the basic emitters in an incandescent source may be considered to add up to give a composite waveform unique to that source at each instant of time; the waveform varies from instant to instant. If the source is compact enough, the light may be split into more than one beam, and each beam will then instantaneously carry an identical waveform. The beams may, therefore, be matched with one another to interfere continuously. A compact source is one which subtends a small angle at the point of interest, so that light from *all* the atomic emitters is gathered into each beam. Lamps used in microscopy are small enough to be considered compact in this sense. The sun also acts as a coherent source, because it is far enough away to be regarded as compact.

Thus anything we have said or will be saying about coherency or interference effects of various kinds is valid for light emanating from a single compact source but cannot be applied to beams which come from separate sources.

If the slit which limits the width of the beam in Fig. 3 is made narrower, the sideways spread of the beam will become of increasing importance. Detailed analysis of the phases of wavelets arriving at various points shows that there will be regions where all the secondary wavelets cancel, giving darkness, and others where a majority reinforce, giving a degree of brightness. The effect of this will be for the beam to emerge from the narrow slit as an undeviated or direct beam (as for a wide slit) accompanied by a series of deviated or diffracted beams, schematically shown in Fig. 5 (*b*). Making the slit still narrower causes the diffracted beams to spread over a wider sector (Fig. 5 (*c*)) and more of the total light is present in these diffracted beams than in Fig. 5 (*b*). The direct beam is sometimes called the zero-order beam and the diffracted beams first-order, second-order beams, etc., counting out from the central beam. If the slit is replaced by a pinhole, the outer beams become annular (in section, that is, they are hollow cones) and the image of the pinhole is seen as a bright disc surrounded by a series of concentric rings of brightness and darkness; this is known as the *Airy disc*.

If the intensity of illumination across such a disc, or across the beams in Fig. 5 (*b*), is plotted against distance from the centre of the direct beam, a characteristic curve of the kind shown in Fig. 6 (*b*) is obtained. It will be seen that the intensity follows a smooth curve, rather than dropping abruptly to zero at the edge of a beam. As indicated above, this result might be expected from consideration of the probable variety of phases of wavelets arriving at a given point; only occasional places will have geometrical relationships which result in a majority of the wavelets being in phase (maximum brightness) or in half opposing the other (darkness). Fig. 6 (*c*) gives the intensity curve appropriate to the narrower slit in Fig. 5 (*c*).

The diffraction pattern from a given slit will clearly depend upon the wavelength of the light employed, for it is this which determines the distance from the elementary sources to points where they reinforce or cancel. This means that the diffracted beams are more spread out for a long wavelength (red) than for a short one (blue), and Figs. 5 (*a*) and

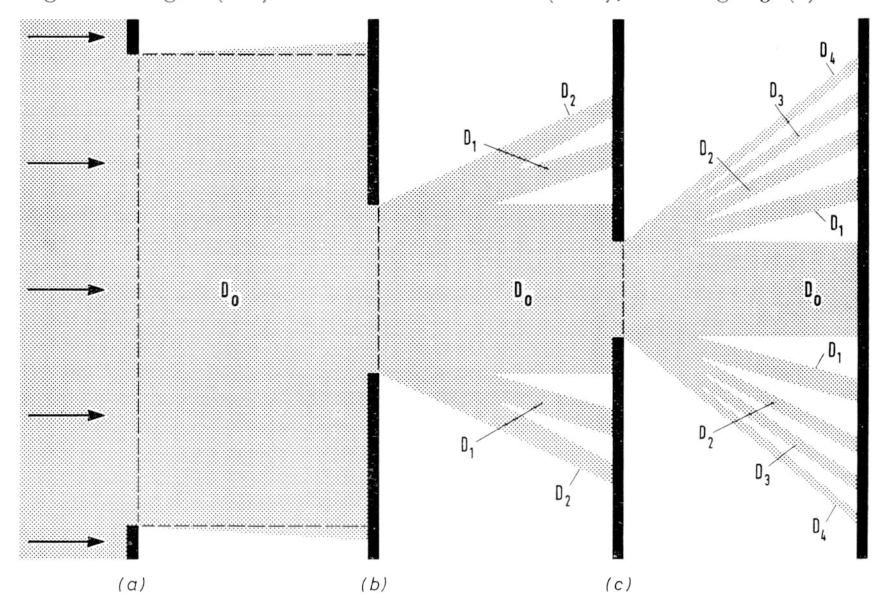

Fig. 5. Limiting a beam by successively narrower slits.
(*a*) A wide slit passes a beam D_0 which is essentially parallel.
(*b*) A sufficiently narrow slit causes diffraction: light is bent into diffracted beams D_1 and D_2, close to the central beam D_0.
(*c*) A narrower slit causes more marked diffraction: a greater proportion of the beam now appears in the widely dispersed beams D_1, D_2, etc.

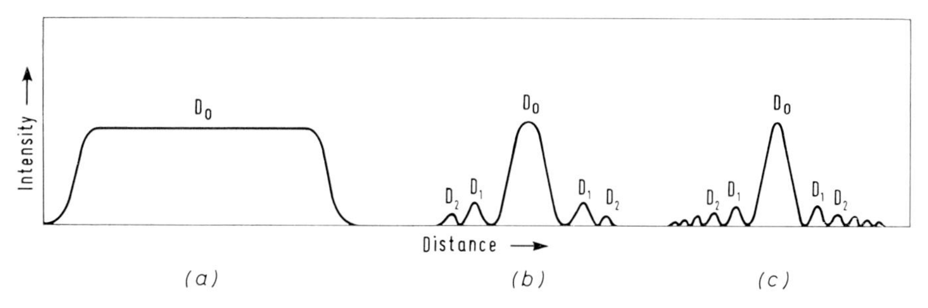

Fig. 6. Intensity distribution for beams diffracted by a narrow slit; diagrams (*a*), (*b*), (*c*) correspond schematically to Figs. 5 (*a*), (*b*) and (*c*) respectively.

5 (*b*), 6 (*a*) and 6 (*b*) could be taken to represent this effect if the slit were assumed to remain unchanged.

One other feature of diffraction is of importance to the consideration of the theory of the microscope. This is that the effect of a single slit can be reinforced by that of other similar slits evenly spaced, with the spacing equal to the width of the slit, to form a *grating*. When illuminated by a parallel beam of coherent light, each slit produces direct and diffracted beams which superimpose to form a pattern, identical with that from a single slit, but of increased intensity.

Abbe's Theory of the Microscope

Abbe* suggested that the formation of an image by the microscope could be understood by considering the object to be a complex of simple diffracting points, like pinholes. The physical processes leading to the final image can then be followed in terms of interference between diffracted beams which pass through the microscope's objective.

The theory is given here for transmitted light because this simplifies the diagram by avoiding doubling rays back on their paths, as would be necessary for a diagram based on the reflecting or metallurgical microscope.

In Fig. 7 the object GG has detail which is represented by a complex array of pinholes P_1, P_2, . . . P_n (not shown in the diagram). Consider one of these, P_j. When illuminated by a suitably collimated† beam of light from the source S, P_j diffracts light in the manner described in the previous section. The microscope objective O collects the direct beam as well as a number of the diffracted beams (as shown in Fig. 7); clearly, if

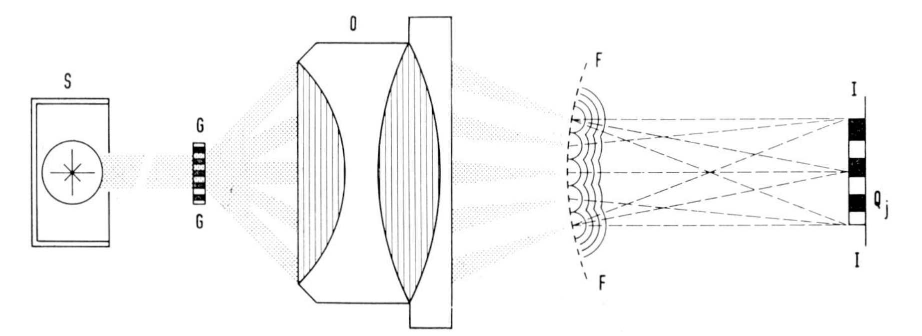

Fig. 7. The Abbe theory of image formation by the microscope. The object GG has detail which diffracts light; the microscope objective focuses the diffracted beams to form a series of images of the source S in the surface FF. These images act as subsidiary sources and give rise, by interference, to reconstruction of detail in the image plane II.

we wish the objective to pass on to the image all the information contained in the light after passing through P_j, it should collect *all* of the diffracted light. (The effect of failing to do so will be discussed later.) After passing through the objective, the direct and diffracted beams will each be brought to a focus at the rear focal plane FF of the objective to form a series of images of the source S (see Fig. 7). These images can be seen if the

* Abbe was a German optical physicist (1840–1905) associated with the rise of the German firm of Carl Zeiss to pre-eminence as manufacturers of optical equipment.

† Collimated means "made parallel".

object approximates to a simple grating—such as would be obtained if a polished specimen were abraded unidirectionally on fine diamond abrasive a few times. This specimen is brought into focus, then the eyepiece of the microscope removed and a "telemicroscope" (such as is used in setting up the phase-contrast microscope, q.v.) is substituted for the eyepiece. When this is focused on the rear focal plane of the objective the direct and diffracted images of the source will be seen; they are usually blurred into a streak with a bright centre because the spacing of the scratches is not uniform (Fig. 8).

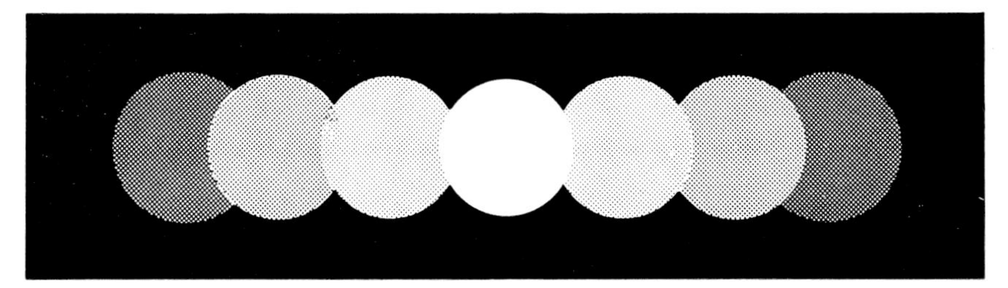

Fig. 8 (schematic). A band of images of the source produced by a microscope objective focused on a specimen with simple microscopic detail—namely parallel scratches. These images constitute the subsidiary sources in the Abbe Theory (Fig. 7).

Abbe considered these images of the light source to be Huygenian subsidiary sources, sending out wavelets which form the image by their destructive interference in the image plane II at the point Q_j. The subsidiary sources (Fig. 7) will be on an arc of a circle centred on Q_j, so that all the subsidiary wavelets will reach Q_j in phase and reinforce to give maximum brightness at Q_j. Around Q_j there will be rings of darkness and brightness, formed by interference; thus, the entire "image" of the pinhole P_j will have an intensity distribution similar to the diffraction pattern previously described (Fig. 6 (*a*)). The process is similar to that illustrated in Fig. 4. The bright central disc is the image of the pinhole and the bright surrounding rings may be regarded as a spurious blurring or broadening of this image.

The first point to notice about this theory is that it makes collection of at least two of the family of beams a necessary requirement for the formation of any image at all: without two beams there can be no interference. The second conclusion which can be drawn is that, ideally, all the beams diffracted by the object should be collected by the objective and passed on to interfere in the image plane. However, since the amount of light diffracted into the beams of higher order rapidly becomes very small as the number of the order increases (see Fig. 6) neglect of the beams of higher order does not in practice seriously affect the image. A third point is less obvious and can, in fact, only be deduced mathematically: this is that the sharpness and intensity of the central bright disc—the image we are examining—also depends on collecting an optimum number of the diffracted beams, as well as the direct beam.

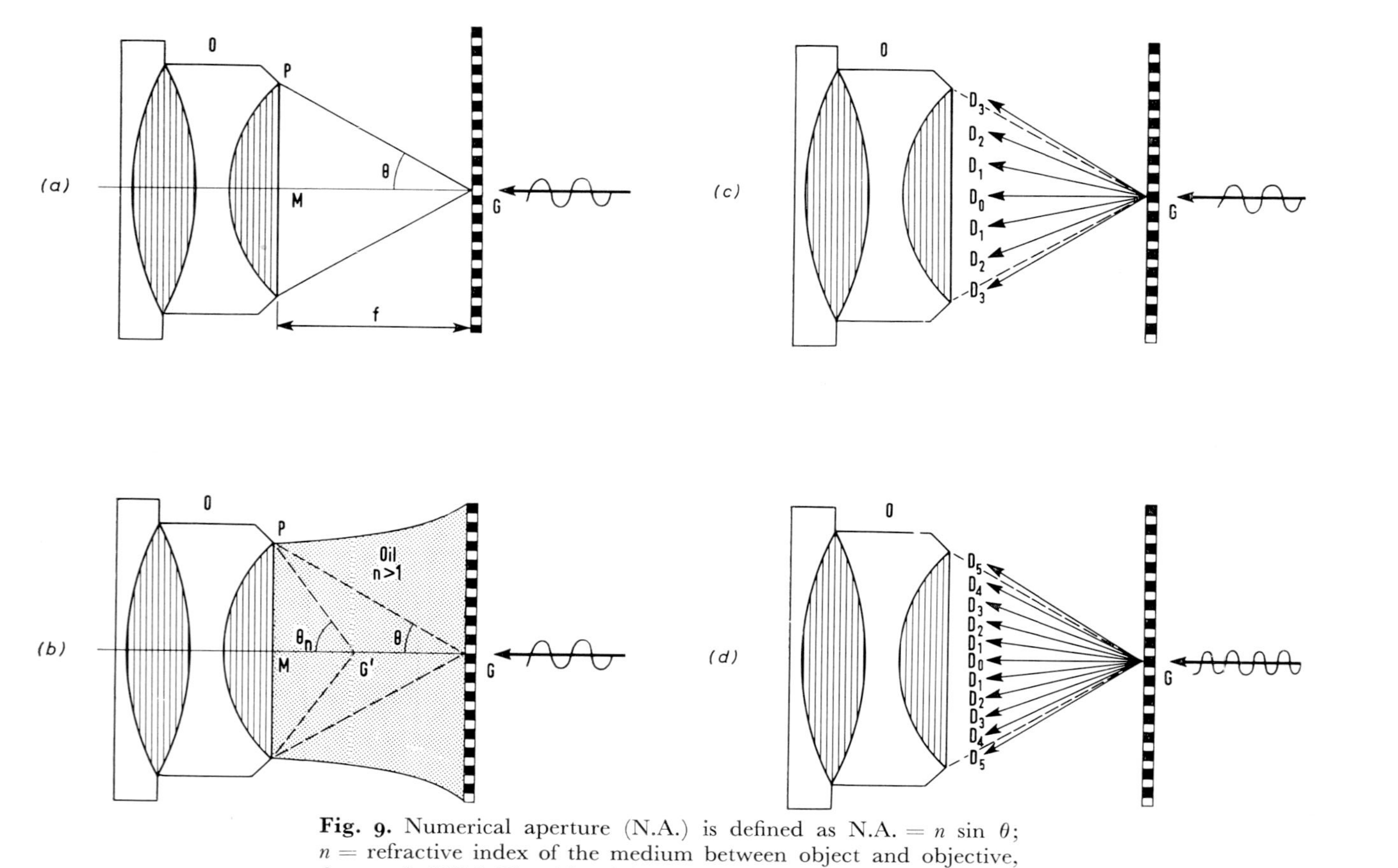

Fig. 9. Numerical aperture (N.A.) is defined as N.A. $= n \sin \theta$; $n =$ refractive index of the medium between object and objective, $\theta =$ angle PGM. The object is in focus (distance f from the objective). (*a*) N.A. in air, $n = 1$. (*b*) N.A. with oil immersion, $n > 1$. (*c*) and (*d*) show the effect of changing the wavelength upon the ability of an objective of given N.A. to collect the diffracted beams D_1, D_2, etc., from an object.

Numerical Aperture

The ability of the objective to gather diffracted beams is clearly a matter of its size in relation to the focal distance. This attribute of an objective is so important that it is specially designated: it is called the numerical aperture (N.A.) of the objective. N.A. is measured by the sine of half the angle subtended, when the object is in focus, at an object point by the light-gathering front component of the lens, as illustrated in Fig. 9 (*a*).

From this it will be seen that, for a "dry" objective in air (remembering that the refractive index n is equal to unity for air)

$$\text{N.A.} = 1 \times \sin \theta \qquad (1)$$

If the space between the lens and the objective is filled with a transparent medium of refractive index greater than unity the N.A. will be increased, because the rays from the object point will now appear to diverge from G' in Fig. 9 (*b*) and so give an effective increase in θ. In this general case we have

$$\text{N.A.} = n \sin \theta \qquad (2)$$

It will be seen in Figs. 9 (*c*) and (*d*) that decreasing the wavelength of the light enables an objective of given N.A. to collect more of the diffracted rays from the object.

Resolution

We are now in a position to examine the question posed earlier, namely what determines the resolution of a microscope?

The criterion for resolution was originally derived by Rayleigh* and modified by Abbe. Suppose it is desired to form an image of an object which consists of two pinholes P_1 and P_2; how close can we bring these and yet obtain an image which shows two spots Q_1 and Q_2 which are clearly separated? From the discussion of the image of a single pinhole, it follows that the intensity distribution of light in the image will be shown as in Fig. 10 (*a*) when P_1 and P_2 are well separated. As they are brought closer (Fig. 10 (*b*)) the diffracted light from the two sources will begin to overlap, but the undeviated beams will remain clearly separate. On still closer approach these will firstly overlap and then merge (Figs. 10 (*c*) and 10 (*d*)). Rayleigh suggested that it should be possible to distinguish the images Q_1 and Q_2 as those of separate objects if, as in Fig. 10 (*c*), the centres of the central bright discs come directly over the centres of the first dark rings. This may appear to be an arbitrary criterion, but it has proved to be a practicable one.

We have seen that the sharpness of the bright central disc and the radius r of the dark ring both depend upon the N.A. of the objective and upon the wavelength of the light used. Specifically, the smallest value of r and the sharpest central intensity are obtained when as much diffracted

* Rayleigh was a British physicist of wide interests (1840–1919).

light as possible is gathered by the objective. This is favoured by a large numerical aperture, (N.A.), and a short wavelength, λ. Thus

$$r \propto \lambda/(\mathrm{N.A.})_1 \tag{3}$$

The term $(\mathrm{N.A.})_1$ has been written here to denote the numerical aperture referred to the image space, i.e. the space between the objective and the image.

There is a general expression due to Lagrange (see Shillaber,[5] p. 189)

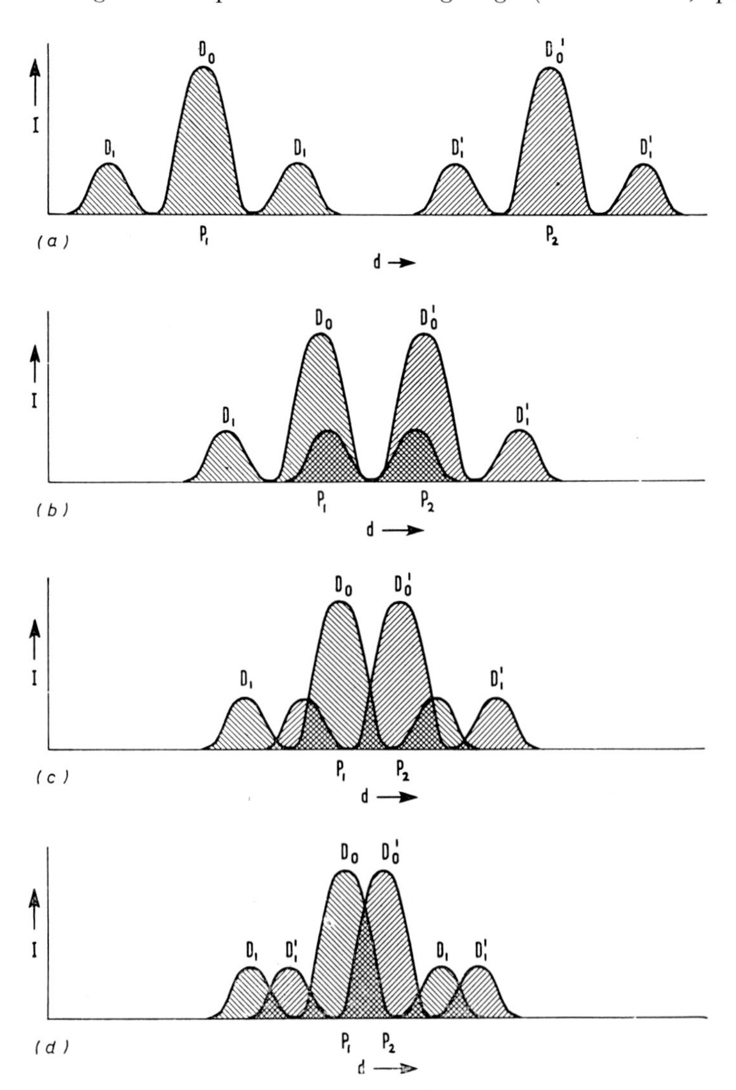

Fig. 10. Intensity distributions for images of two points (pinholes) P_1 and P_2 in an object. As P_1 and P_2 become more closely spaced the separate images in (*a*) begin to overlap (*b*), cannot be resolved after stage (*c*), and become completely merged in (*d*). Stage (*c*) defines the limit of resolution, the maxima of the direct beams D_0 and D'_0 falling above the first minima.

which relates $(N.A.)_1$ to the usual numerical aperture (N.A.) through the magnification M, namely

$$(N.A.)/(N.A.)_1 = M \tag{4}$$

Hence we can write equation (3) as

$$r \propto M\lambda/(N.A.) \tag{5}$$

or

$$r/M \propto \lambda/(N.A.) \tag{6}$$

From the discussion of Fig. 10 we have seen that Rayleigh's criterion is that we cannot resolve two points in the image which are spaced closer than a distance $2r$; it follows that the corresponding distance in the object is $2r/M$ and this distance is defined as the resolution R. In other words, if two points in the object are separated by a distance smaller than R we shall not be able to resolve them in the image. Hence we can rewrite equation (6) to define R:

$$R \propto \lambda/(N.A.) \tag{7}$$

or

$$R = K\lambda/2(N.A.) \tag{8}$$

where K is a constant.

Various estimates of the constant K have been made. Abbe used a theory based on the image from a diffraction grating and so found $K = 1$. Rayleigh used the approach we have followed and considered two point sources in the object. Then $K = 1{\cdot}22$ if the objective is filled with light by an appropriate condenser system. If the light is limited by an aperture stop (or any maladjustment of the illuminating system) K takes other, lower, values. The reason why the constant is generally written so that the denominator in equation (8) contains the factor 2 is that the equation can then be taken to include the illuminating condition. Maximum resolution is obtained when the N.A. of the illuminating condenser is numerically equal to that of the objective, so that the effective factor for the system is (N.A.) + (N.A.) = 2 (N.A.). We shall see that this also applies to metallurgical microscopes, because the objectives in these act as illuminating condensers.

Sometimes the term *resolving power* is used for the reciprocal of the resolution R, and some writers use the two terms interchangeably; it seems best to keep to one clearly defined term; the logical choice is then that of resolution, for this indicates directly the distance between details which can be resolved.

IMPLICATIONS

There are some implications of equation (8) which are worth noting. It has already been stated that a gain in resolution is achieved by shortening the wavelength of the light used, such as changing from green to blue

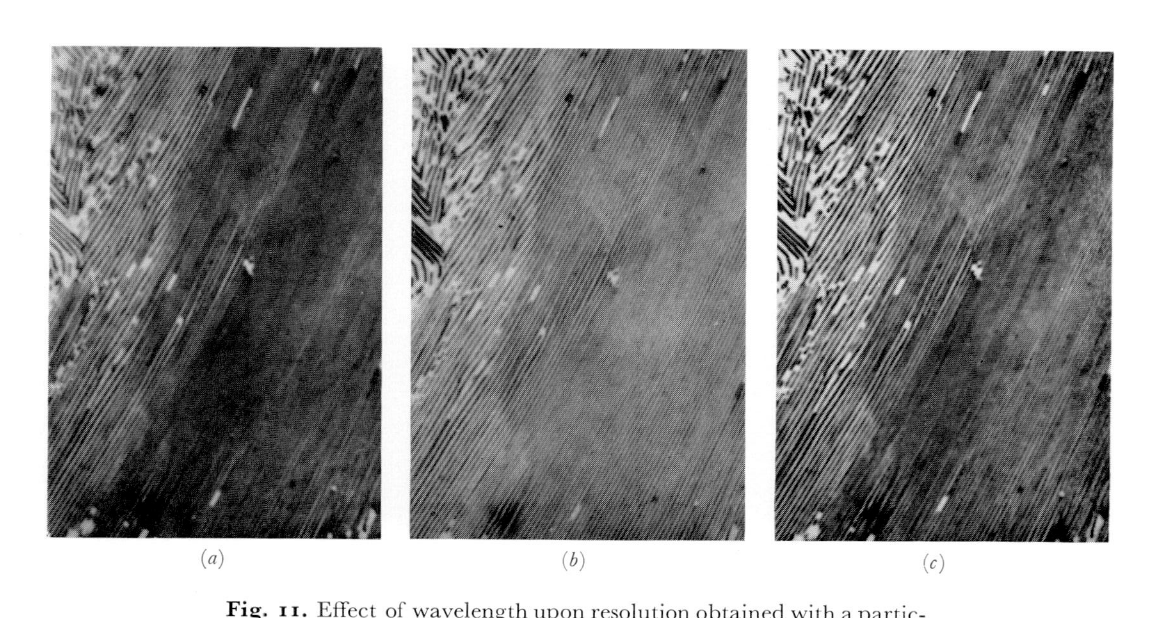

(*a*) (*b*) (*c*)

Fig. 11. Effect of wavelength upon resolution obtained with a particular objective (N.A. = 1·4). Pearlite structure, × 2000.

(*a*) Partially resolved with red light (~ 6800 Å).
(*b*) More completely resolved with green light (5500 Å).
(*c*) Even more completely resolved with blue light (~ 4700 Å).

(Reduced to $\frac{9}{10}$ in reproduction)

light, i.e. from a wavelength of 5500 Å to 4700 Å. The improvement for this particular case is not great; for example, with N.A. = 1·4, from 0·24 to 0·20 μm. However, as shown in Fig. 11, this improvement may be critical for a very common metallurgical structure—normalized steel. A further improvement can be brought about by changing to wavelengths beyond the visible spectrum, for example by the use of ultra-violet light. Ultra-violet microscopes have been made, but they are expensive because

(*a*) (*b*)

Fig. 12. Effect of N.A. upon resolution. Pearlite structure, ×1500. (*a*) N.A. = 1·25 (fluorite); (*b*) N.A. = 0·65 (achromat).

quartz lenses are required to avoid absorption of the ultra-violet light by glass components; moreover, there are difficulties in focusing the ultra-violet light, which is not visible to the eye.

There was a time, 15–20 years ago, when the small but costly gain in resolution was considered worth while; today, however, there are much greater gains to be had as a result of very short wavelengths arising from the use of electron beams. The wavelength of an electron beam depends on the accelerating voltage, but typically may be only 0·1 Å, thus increasing resolution by $5000/0{\cdot}1 = 5 \times 10^4$ compared with a microscope using green light. The potential improvement in resolution is, therefore, very great but it is limited by design problems; however, electron microscopes now commonly give a resolution of better than 10 Å. This enormous advantage is partially offset by difficulties associated with the fact that electron microscopes are almost always used in transmission (only very oblique incidence being an alternative possibility).

Since metals are opaque to electrons unless they are prepared as very thin specimens, electron microscopy requires either specially prepared *thin foils* or *replicas* of the specimen surfaces. We shall discuss replicas for optical microscopy in Chapter 9. Excellent texts on electron microscopy are available (see *Select Bibliography*, p. 200), and these should be consulted for further information on the operation of electron microscopes and on their advantages and limitations.

A second point arising from equation (8) is that the gain in resolution which can result from increasing the N.A. is a little less than a micron. With the N.A. of a typical low-magnification objective (say a 16 mm lens) the resolution is 1·1 μm and with a good oil-immersion lens it would be of the order of 0·2 μm. In metallography it is rather rare for the resolution itself to be the limitation: contrast, flatness of field and other aspects of image quality are often of overwhelming importance. It so happens that these attributes do not all improve with lenses of high N.A., so that the selection of the best objective for a particular purpose is not always solved by simply choosing on the basis of resolution. This will be discussed in greater detail in the chapter dealing with the use of the microscope. However, pearlite can once again be used to show that the choice of N.A. can be critical, see Fig. 12.

Perception Limit

A third point of importance, mentioned briefly at the beginning of this chapter, is that the equation for resolution deals with the separation between two small features, not with perception of an isolated small feature. This is often the important aspect of metallurgical specimens and therefore deserves some attention. Reference to Fig. 6 will show that the image of an isolated pinhole or disc is broadened by a small N.A. or a long wavelength of light, but this of itself is not detrimental to the *perception limit* of the image of the feature. The more important role of N.A. in this context is that a high N.A. enables more light to be diffracted into the central bright beam, and so aids perception through improved contrast. A full treatment of perception is extremely complex. Françon[4] shows that perception is also dependent upon the aperture of the illuminating diaphragm (see Chapter 3). The perception limit for a small black disc on a white ground is given by a radius r_p where

$$r_p = 0{\cdot}08\lambda/2n \sin \theta \tag{9}$$

For a thin black line it is a width w where

$$w = 0{\cdot}01\lambda/2n \sin \theta \tag{10}$$

However, the perception limit of a small white disc or a thin white streak on a black ground depends solely upon the luminous flux from the feature, i.e. upon the total light reflected per cm^2. This is an important point which bears upon the use of dark-ground illumination (q.v.).

Effect of Focusing upon Image Sharpness

We have so far assumed that the microscope is precisely focused; that is, the images discussed have been those in the true image plane. The true image plane is such that the Airy disc has the theoretical distribution of light in the direct and diffracted beams (Fig. 6). It is clear from the Abbe theory that the destructive interference between waves from the subsidiary sources will produce different (but related) distributions of light at points in planes close to the true image plane. The effect, in general, is to reduce the intensity of the direct light, and to diffract more light into the luminous rings. If the objective is "perfect", that is, it is completely corrected for aberrations, the diffraction pattern obtained will be symmetrical either side of true focus. This fact can be used to test a microscope objective for correction, by using a pinhole in a vacuum-coated film of metal as a test object and observing the degree of symmetry close to true focus.

The remarks just made about symmetry of the diffraction pattern apply to the image from a planar object. In transmission microscopy, preparation by sectioning leads to such a specimen, but in metallography the surface examined is rarely planar. Surface roughness of some kind is generally present, having been introduced by etching, deformation or even the polishing process itself. Since only one level of the specimen can be in true focus at a time, these changes in level have the effect of varying the degree of departure from true focus and thus modifying the image. The importance of the modifications introduced depends upon factors such as the N.A. of the lens, its focal length and the setting of the aperture stop. Some of these points will be taken up in later sections (Chapter 5). Perhaps the most important consequence of this out-of-focus effect is in focusing itself, for the changes either side of focus allow true focus of a hitherto unknown detail to be recognized when the focusing control is moved fairly rapidly to-and-fro through the focus position.

REFERENCES

1 F. A. Jenkins and H. E. White, *Fundamentals of Optics*, New York, 1957 (3rd edn) (McGraw-Hill).
2 B. B. Rossi, *Optics*, Reading (Mass.), 1957 (Addison-Wesley).
3 L. C. Martin, *Technical Optics*, 2nd edn., Vols. 1 and 2, London, 1966 (Pitman).
4 M. Françon, *Progress in Microscopy*, Oxford, 1961 (Pergamon).
5 C. S. Shillaber, *Photomicrography in Theory and Practice*, New York, 1944 (Wiley).

3 The Metallurgical Microscope

In Fig. 13 (*a*) the essential components of a "linear" or optical bench form of metallurgical microscope are shown schematically. The arrangement illustrated was basic to most first-class microscopes (such as that shown in Fig. 13 (*b*)) until recently, when it was generally replaced by a

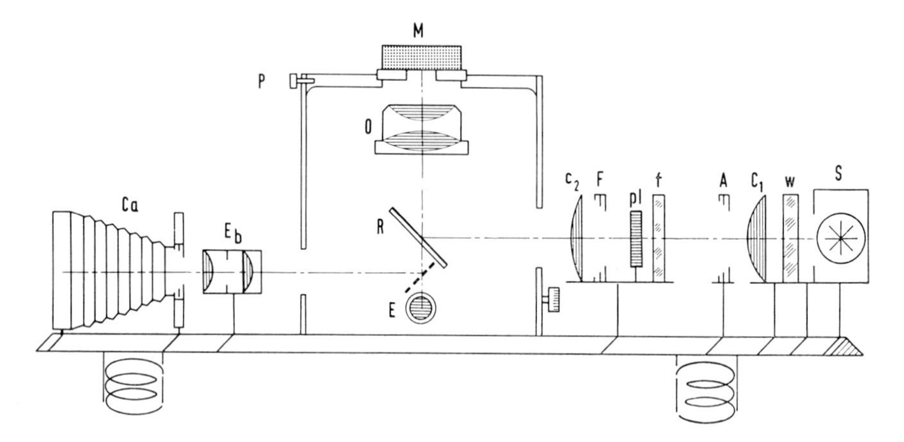

Fig. 13(*a*). Schematic representation of the essential components (capital letters) of a large, photographic, metallurgical microscope—a *metallograph*. Some ancillary equipment is also included; for key to lettering, see text.

more compact design, as illustrated in Fig. 14. When these large modern microscopes are equipped for photography, they are often called "metallographs". Smaller microscopes, which may also have attachments for taking photographs, are called bench microscopes because they are placed on a bench or table for use; one of these is illustrated in Fig. 15.

The essential components in Fig. 13 (*a*) are labelled with capital letters, ancillary equipment with small letters. Illumination is provided by the source S which is focused by a condenser lens C_1; the size of the beam is controlled by a diaphragm, the aperture stop A. A second diaphragm F is called the field stop, for it limits the field of view on the specimen. The beam is then reflected through 90° by an illuminator R and passes through the objective lens O, which acts as a condenser and focuses the beam on to the specimen M which is held on the stage P. Light from the specimen is reflected back first through the objective, which now acts as a magnifying lens, and then through the reflector which is so arranged

Fig. 13(*b*). A metallograph conforming to the scheme shown in (*a*). (Note that this is *not* the current model; see Fig. 14.)

that part of the light continues straight on to the eye (or eyes) through a secondary magnifying system called the eyepiece or ocular, E.

The principal requirements of each of these components, and the usual practical methods of meeting them, are outlined in the following paragraphs.

Fig. 14. A metallograph which incorporates a full range of accessories, viz. phase-contrast, polarized light, automatic exposure, etc.

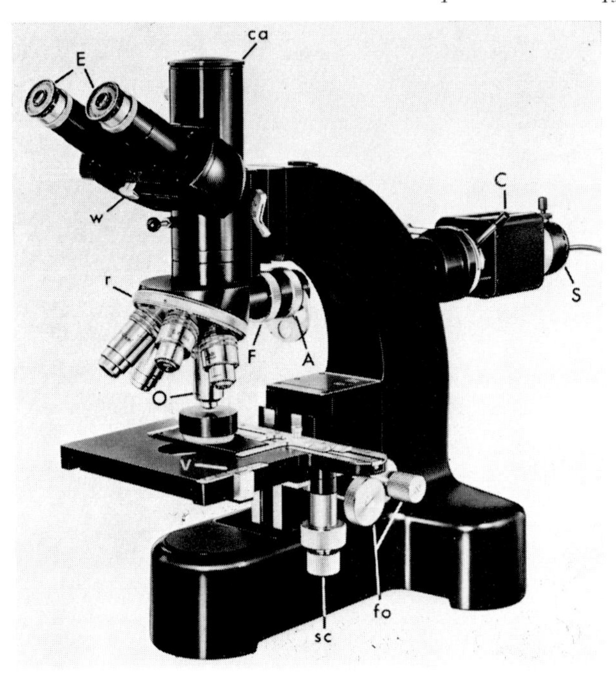

Fig. 15. A research bench-type microscope.

A = aperture stop	ca = tube for camera
C = condenser lens movement	fo = focusing controls
E = eyepieces (binocular vision)	r = rotating turret
F = field stop	sc = stage controls
O = objective	v = vernier for stage movement
S = source	w = width control for eyepieces

Source

The source of light should be bright, steady and compact. It needs to be bright because, even with a highly reflecting specimen, it is not usual to finish with more than 20–25 per cent of the original intensity of illumination at the eyepiece. A low intensity, which might be sufficient for visual examination, could give difficulties in focusing for photography or in display for instructional purposes. Also, the resulting length of photographic exposures may increase the risk of vibration, causing lack of sharpness in the photographic record. The source has to be compact in order to approximate to the point source which is theoretically required to obtain even, coherent illumination. The steadiness of the source is a matter of comfort in viewing and of reproducibility in photography.

Tungsten filament lamps of comparatively low intensity are often used for visual examination, and special tungsten filaments in the form of strips are available for photography, but most microscopes use some kind of arc lamp for this. The "Point-o-lite" is a proprietary lamp using a tungsten ball in an evacuated bulb as the luminous electrode in an arc; it gives a brilliant, steady light but one which is not as bright as other arc lamps to be described. Carbon arcs were a standard fitting until recently; their main disadvantages are that only rarely does a really satisfactory mechanism seem to have made for moving the carbon rods to keep pace with their consumption and also that the centre of the arc tends to wander, despite the use of cored rods of carbon. High-pressure mercury arcs are very bright and have long-term stability, but tend to flicker irritatingly. They have variable and unpredictable lives, are relatively costly to replace and they also have minor disadvantages, in that they take about ten minutes to reach full intensity and cannot be switched on again whilst still hot. Since they emit ultra-violet light, provision should be made to shield operators, especially their eyes, from the direct light. A similar precaution applies to xenon lamps, which are becoming increasingly popular; they are bright and steady and strike immediately to give full intensity when both hot and cold. They are also relatively expensive. Mercury and xenon lamps require special chokes or transformers for ballasting the electrical circuit when striking the arc, as well as requiring relatively complex equipment for switching the ballast in and out of circuit.

Condenser C_1

This serves two functions; it collimates the light from the source, to assist in providing a beam of the requisite size, and it also ensures even illumination of the specimen, by being incorporated into a system of adjustment which will be described under the heading of critical illumination.

Aperture Stop

The aperture stop consists of an iris diaphragm placed close to the condenser C_1. Its function is to control the amount of light by limiting the diameter of the beam; thus, full use can be made of the potentialities of the objective lens, without flooding the microscope with light and causing degradation of the image by stray reflections. Under special circumstances, which will be discussed in later sections, this stop may be closed down to give a very narrow beam.

Field Stop

This is also an iris diaphragm, placed so that the source, or more usually the condenser C_1, is imaged on it. By means of another condenser, C_2, and the objective lens itself the field stop is then imaged on to the surface of the specimen. Adjustment of this stop, therefore, controls the area (or field) of the specimen which is illuminated.

Illuminator

The illuminator can take several forms. For most work, and in the majority of microscopes, it consists of a thin plate or disc of plane glass (sometimes referred to as a "pellicle") placed in the illuminating beam with its normal at 45° to the beam. Because of its thinness, distortion of the image by multiple reflections is kept to a minimum. Light is thus reflected through a right angle into the objective lens. To increase the efficiency of the device, the front surface of the glass is usually partially silvered and, to prevent unwanted reflection from the back surface, it may be coated with an absorbing material (or "bloomed") on this surface. Because part of the original beam also passes straight on through the glass, and a similar division occurs again after the beam has returned through the objective and been reflected at the specimen, this glass disc is often called a beam-splitter. It necessarily restricts the intensity of the light passed on to the ocular to less than 25 per cent of that at the aperture stop.

An alternative illuminator consists of a small right-angled glass prism. This gives a more intense illumination, since the beam is totally reflected by it. However, in order to allow the beam to return to the eyepiece, the prism has to be placed to reflect only half of the original area of the beam. This means that half of the objective lens is being used as a condenser, half as a magnifier, and this necessarily reduces its resolution. A further consequence of the prism being placed off-centre is that the illumination on the specimen is slightly oblique; this may or may not be desirable.

Some microscopes have another alternative illuminator consisting of a polished metal tongue, sector-shaped, which gives oblique illumination of the same kind as the prism. The only use for the metal-tongue illuminator today is in a system called "light-cut illumination" (see p. 99).

One widely used microscope (Bausch and Lomb) employs an illuminator consisting of a pair of calcite prisms cemented together in such a way as to reflect all of the light from the specimen into the eyepiece. At the same time, the illuminating beam is rendered plane polarized (see Chapter 7). The illuminating beam is normal to the specimen surface and yet gives a more intense final image than does a prism illuminator. This device, known as a Foster prism, is fitted only to this one make of microscope.

Objective Lens

The objective lens is the most important and critical component of the microscope; in a sense, of course, it *is* the microscope, the other parts being ancillary to it and designed to render its use efficient.

All objectives consist of a number of glass lenses, sometimes combined with lenses of fluorite (naturally-occurring calcium fluoride) to form a compound lens which gives the desired magnification, which is usually in the range of × 5 to × 160. A cut-away representation of a typical microscope lens is shown in Fig. 16. In order to obtain adequate resolution, the corresponding range of values of numerical aperture is from 0·2 to 0·9 (dry) or 1·4 (oil immersion). The upper values are set by the fact that an angular aperture of $2\theta = 145°$ is as large as is generally practicable, for the lens curvatures involved are then as small as can reasonably be attained.

The various objectives are often mounted so that they can slip into an appropriate housing on inverted-type microscopes; some are held by magnetic chucks. Particularly with bench microscopes, the objectives may be mounted on a rotating turret which can bring each lens into position quickly, and it is increasingly the case that the objectives are designed to be parfocal. This term implies that the specimen remains in focus when the objectives are changed, despite their different working

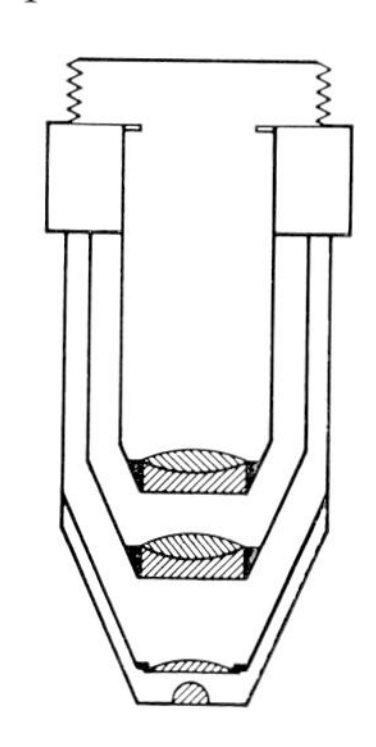

Fig. 16. A typical microscope objective, and cut-away representation.

distances. The working distance, as its name implies, is the clearance between the front of the objective and the specimen, when the microscope is focused.

In designing an objective lens, especially for the highest resolution, some degree of compromise has to be tolerated, for there are several features other than resolution which affect the performance of the lens. These are grouped under the heading of aberrations. Since correction of aberrations is part of lens design and not something the microscopist can control except by purchase, fuller details of aberrations have been relegated to the Appendix of this chapter and only a summary is given here. *Chromatic aberration* refers to the fact that (with white light) an object point will be imaged for different wavelengths in different planes; this is known as longitudinal chromatic aberration. Various wavelengths may also give rise to various magnifications, an effect which is called lateral chromatic aberration. Other aberrations are grouped together under the heading of *spherical aberration*; this includes the blurring of the image of an object point which arises from the changes in focus from the centre to the outside edge of a lens. Other defects in the shape of a lens can give rise to *astigmatism*, whereby points become drawn out into lines in the image, and *coma*, which produces comet-like tails on images of points. Of great importance in many metallurgical applications is *flatness of field*; this is the ability of the lens to image a sufficiently large area of the specimen sharply in one plane without serious intrusion of the other aberrations.

Corrections for all the aberrations are made by using, as unit components in the objective, lenses of appropriate curvature in materials of various refractive indices; these are either optical glasses or fluorite. The number of aberrations to be corrected for and the range of wavelengths in white light are such that the glasses available cannot provide full correction, and hence there is the need to compromise. Spherical aberration, astigmatism and coma are almost always well corrected, especially in objectives of high resolution.

Corrections for longitudinal chromatic aberration are made to various

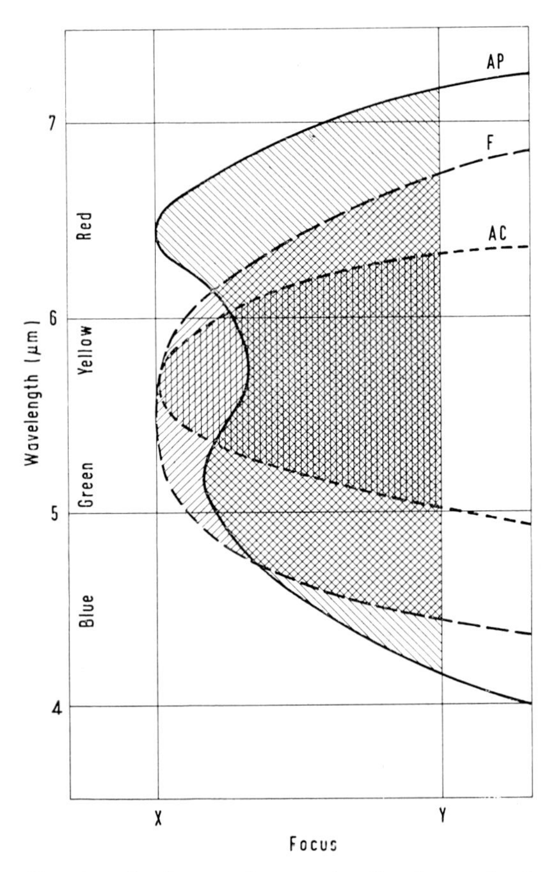

Fig. 17. Showing the degree of correction for chromatic aberration in objective lenses of the achromatic (line AC), fluorite (line F) and apochromatic (line AP) types. XY represents the useful depth of focus attainable.

degrees by different classes of lenses: non-achromatic, achromatic, semi-apochromatic (or fluorite) and apochromatic. For metallurgical work, non-achromatic lenses are not used. The others will now be described briefly.

Achromatic lenses are corrected to bring all rays in a limited region of the spectrum to essentially the same focus. The region chosen is the yellow-green with wavelengths between approximately 5000 and 6300 Å (see Fig. 17). In practice the chromatic correction is made for two specific wavelengths in this range and correction for spherical aberration is also made for one wavelength in this part of the spectrum. With the higher powers it is usual to include fluorite components and hence some of the achromat series may be called *fluorites* or *semi-apochromats*. These commonly give good chromatic correction over a range of wavelengths extending between 4500 Å and 6500 Å. When corrections are made chromatically for three colours and spherically for two, the lenses are known as *apochromats*. The range of wavelengths brought to the same focus is then approximately 4200–7200 Å (Fig. 17). However, in order to obtain optimum correction, apochromats must be used with appropriate compensating oculars (q.v.) because they are, in fact, undercorrected for colour in order to permit the fuller correction for spherical aberrations. This is an important point and means that there are advantages other than colour correction which accrue from the use of apochromats.

Specimen Stage

A rigid platform to hold the specimen in the focal plane of the objective is also a necessary part of the microscope system. In the larger microscopes the stage is often *inverted**—that is, the objective is below the stage and the specimen is placed on one of a series of annular discs so that it can be viewed whilst it is resting on the disc. In the other type of microscope the specimen, with its polished face upwards, is placed or held on a stage, the objective lens being above it. Both kinds of stage are fitted with guides so that the specimen may be moved smoothly in two orthogonal directions. It is an advantage if these movements can be controlled by slow-motion or pressure devices actuated by one hand, leaving the other hand free for focusing. It is also useful to have the movement of the stage measured on vernier scales. For rapid scanning of a specimen some microscopes have stages which can be moved directly, but slowly and positively, by pressure from the fingers of one hand. Stages are often also capable of rotation about an axis which can be made to coincide with the optical axis of the objective, the amount of rotation being measured on a circular scale; this feature is of particular use when examining specimens in polarized light, but is also invaluable when "composing" photographs of structures.

Focusing is usually effected by movements of the stage, on a coarse scale by rack and pinion and on a fine scale by an auxiliary screw thread. It is an advantage if the knob giving the fine adjustment is divided to show the movement of the stage in fractions of a millimetre.

Eyepiece or Ocular

After passing back through the beam-splitter R the rays from the object are brought to a focus and viewed by means of an auxiliary magnifier known as the eyepiece or ocular. The purpose of this lens is either to give,

* Sometimes described as a "Le Chatelier"-type stage.

for the eyes, a virtual image or to project the primary image on to a photographic plate; in both cases it serves to increase the initial magnification given by the objective. It may also be used to assist in correction of aberrations in the objective.

There are three principal types of ocular.

(i) *Ramsden or Positive:* As with all oculars, this is a combination of simple lens units; it is a positive lens because it can form a magnified image of an object, just as a simple hand-magnifier does. A feature of the lens is that the rays from the objective are arranged to come to a focus on a diaphragm below the lenses. It is thus possible to place a micrometer or other scale on this diaphragm and use it to calibrate the objective, measure grain sizes, etc. The basic principle of a Ramsden eyepiece is shown in Fig. 18 (*a*).

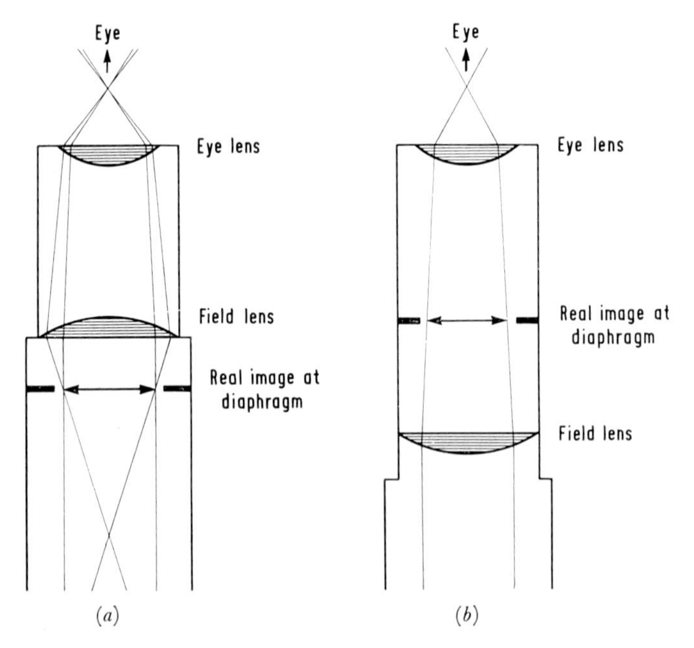

Fig. 18. Disposition of lens components in two principal types of ocular (or eyepiece): (*a*) Ramsden; (*b*) Huygenian.

(ii) *Huygenian or Negative:* This ocular is so constructed that it produces a slight reduction of the primary magnification of the objective and, with the eye, a virtual image some 250 mm from the eye. It is illustrated in Fig. 18 (*b*). It only gives a well-corrected image over the centre of its field. It is suitable for low-power photographic work, but to form a real image on the photographic plate the microscope has to be refocused, changing it from correct focus for the eye.

(iii) *Amplifying:* These lenses should not properly be called oculars, since they are designed to project the image on to a focusing screen or photographic plate; they should more correctly be called projection lenses, but this term seems to be reserved for lenses which project the image to somewhat greater distances (10–15 ft) than those used in photography. Amplifying lenses are of negative type, in that they produce a reduced image when used as a hand magnifier.

COMPENSATING OCULARS

The basic types of ocular may all be modified and classed as compensating oculars. These are designed to correct the residual chromatic aberration of apochromatic objectives. They can usually be distinguished by the fact that a blue ring around the ocular diaphragm is seen when inspecting the field of view. A compensating ocular with an apochromat gives a field of view which is less flat than that of a combination of an achromatic objective and a Huygenian ocular. Many makes of microscope are furnished, therefore, with special oculars which have intermediate corrections for chromatic aberration, so that a flatter field can be obtained with them than with the combinations using fully compensated oculars. Because flatness of field is often of greater importance in metallography than full correction for other defects, use of these intermediate oculars is very common. Various manufacturers offer intermediate oculars designed for use with their own particular microscopes and objectives. It is therefore essential that the specified combination of objective and ocular should be used.

For visual work, *binocular eyepieces** are becoming increasingly used. They consist of an attachment which fits into the eyepiece tube and which contains prisms to split the rays and reflect them into two similar oculars; these can be moved laterally to adjust their spacing to that of the eyes of the operator. A binocular eyepiece gives no advantage from the purely optical point of view, but contributes to a reduction in strain in viewing, particularly for those who find difficulty in defocusing the eye which is not being used with normal monocular viewing. Closing the unused eye is not good practice, because it leads to muscular strain.

EYEPIECE RETICLES OR GRATICULES

Oculars can incorporate graticules of various kinds, so that the scale or figure on the graticule appears superimposed on the image of the specimen. More elaborate eyepieces are available with built-in graticules and means for focusing them sharply by rotation of the components of the ocular nearest to the eye. Further reference to these will be made in the section on quantitative metallography, in connection with the measurement of grain size and with point counting. Even more elaborate eyepieces have a movable vernier scale; these are referred to as *micrometer eyepieces*.

* See also Chapter 9, "Single-objective Stereomicroscopy".

Ancillary Equipment

To complete this survey of the metallurgical microscope the ancillary equipment shown in Fig. 13 (*a*) will be described. Many sources of illumination radiate considerable amounts of infra-red which, when focused, might damage other components of the microscope. To absorb this heat a glass cell containing water or a heat-absorbing glass filter (w) is inserted close to the source.

Provision is also made close to the field stop at (f) for the insertion of of other *filters* of various colours. The eye has its greatest sensitivity in the yellow-green portion of the spectrum, which is a reason for the various degrees of chromatic aberration being centred on this region; the most useful filter is therefore a yellow-green one. This also has an advantage in that it can be used effectively with photographic plates which have not been rendered panchromatic. With an apochromatic objective advantage can be taken of the increased resolution obtainable using a shorter wavelength of light, and blue filters may be used. However, the intensity of the image is often very much lower (some arc lamps having most of the emission in the green region) and this combines with the lower visual sensitivity to make focusing difficult. Translucent diffusing screens are also often supplied; their use should be confined to macroscopic—as distinct from microscopic—work.

For working with polarized light, microscopes may be fitted with a polarizer (pl) close to the field stop and an analyser between the beam-splitter and eyepiece. These will be discussed in the section on the use of polarized light (see Chapter 7).

For photography a prism may be moved to intercept the beam and reflect it into a camera Ca. On a large microscope this camera is generally fitted with a bellows, so that some further magnification can be made of the image from the projection eyepiece E_b. On smaller microscopes the camera is in the form of an attachment which clamps on to a tube which replaces the ocular used for visual examination. This arrangement is usual for recording on 35 mm film, when a conventional 35 mm camera back may be used.

The metallograph described in this chapter is essentially a microscope capable of giving the full range of adjustments and producing photographic records on standard sizes of film or plates. The simpler and less expensive bench microscopes are also widely used, mainly for visual examination of specimens rather than for photography. Their essential features are the same as we have described and must meet the requirements of stability, illumination, etc. which have been listed. Bench microscopes are often arranged with the stage below the objective, so that the specimen rests on it, face up. This means that the specimen must be levelled in some way, to make the optic axis of the microscope normal to the plane of the specimen surface.

Bench microscopes range from very simple types, often known as "student" microscopes, to ones which are virtually metallographs built on the bench-type principle, although these may take smaller sizes of photographs because no bellows extension is provided.

It is always desirable to have a microscope mounted on a table or bench which is as free from vibration as possible; this is essential for good photography, and metallographs usually have anti-vibration devices built into their construction.

Other important auxiliary details of a good microscope are devices for centring the aperture and field stops and for centring the specimen stage. In addition, there are many other pieces of equipment designed to operate special techniques, such as will be described in Chapter 5.

Appendix

Aberrations of Lenses

Aberrations of lenses are of two main classes—spherical and chromatic. Spherical aberrations are those which are present even when monochromatic light is used and comprise spherical aberration itself, coma, astigmatism, curvature of field and distortion. Chromatic aberrations may be lateral or longitudinal and refer to errors in the focusing of rays of different wavelengths.

SPHERICAL ABERRATION

If a lens is ground to have truly spherical surfaces, parallel rays are brought to slightly differing points of focus according to their path through the lens, through its centre, or through its outer zones. The image of a point source at the centre of the field then becomes split into series of overlapping and unfocused images and thus presents a "circle of confusion" instead of a sharp spot. The spherical aberration of a "spherical" positive lens is exactly the opposite of a negative lens having the same radii for its surfaces. Thus two lenses can be combined to form a doublet and eliminate spherical aberration, although the "figuring" required is more complex than this description would indicate, for combining a positive and negative lens of identical curvatures would result in zero magnification. It is also possible to correct for spherical aberration using one lens alone, by selecting appropriate radii for its faces; such a lens is called "aspherical".

It is rare for a cover slip or other glass component to be introduced between the specimen and the objective in metallurgy, but if it is, the correction for spherical aberration is upset.

COMA

Coma is an aberration affecting parts of the image away from the centre. Essentially, it is a result of differences in magnification which arise from the paths of rays which meet the lens at widely differing angles. The image of a point becomes more like a comma—it has a tail, like a comet. To correct for coma the lens surfaces have to be figured so that the sine of the angle made by each incident ray bears a constant ratio to the sine of the angle of the corresponding (emergent) refracted ray.

ASTIGMATISM

This aberration refers to errors which cause points to be imaged as blurred lines (or possibly discs). Literally, the word "astigmat" means "not a point". A lens which is corrected for this aberration is called an "anastigmat". Astigmatic errors may be highly directional: in the eye, for example, they may manifest themselves in such a way that the figures on a clock face vary in clarity. The 12 and 6 may be clear and the numbers fade into blurs for the 3 and 9. As with coma, astigmatism is, in practice, an aberration of the outer zone of the lens, so "stopping it down" by closing the aperture stop reduces astigmatism. This fact has long been used by photographers, who can take this evasive action without the microscopists' concern for resolution to worry them. For microscope objectives, yet another variation from the spherical surface has to be introduced, together with a coupled matching of the refractive indices of the lens components, in order to counter astigmatism.

CURVATURE OF FIELD

Even when the previous three aberrations have been successfully corrected, there remains a serious defect of the image. The sharpest focus for the image of an extended object is not then a plane but a curved surface. For visual examination, this is not of great importance, for the eye can be moved to "follow" the curved image; but it is crucial in photography. The defect can only be overcome in the design of the objective by sacrificing extreme sharpness of all the points in the image. This is not a tolerable solution, for microscopy cannot reveal useful detail if the image is blurred. The solution adopted is to introduce the final correction for curvature of field into the ocular. It is for this reason that the recommended combinations of objectives and oculars *must* be used in photographic work if a flat field is required.

DISTORTION

Oculars are more prone to show this aberration than are objectives. It is a very important form of distortion in metallurgical work and describes the tendency for straight lines, particularly towards the outer edges of the image, to be bowed (see Fig. 19). A familiar example of distortion (caused by somewhat different optics) can be seen on a television screen; the image of, say, a rectangular picture frame is reasonably true at the centre of the screen, but as the zoom lens of the camera enlarges the image it begins to bow out as it approaches the edge of the screen. Distortion is another effect of unequal magnifications being produced by various zones of the lens and is not likely to be found in the central portion of the image.

The metallographer must test various combinations of lenses, using some suitable pattern such as a stage micrometer, and decide which combinations produce tolerable distortion and which cause bowing which is large enough to falsify the record of microstructures.

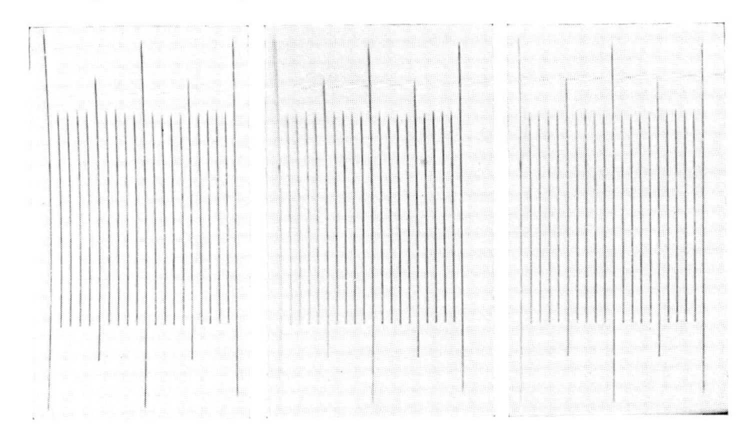

Fig. 19. Distortion is not readily detected in normal micrographs but is shown on photographs of a stage micrometer scale. Distortion is shown in (*a*) and (*b*) and is minimized by the correct choice of ocular in (*c*).
(*a*) 10× Amplifier.
(*b*) 12·5× Compensating.
(*c*) 10× Hyperplan.
Total magnification, ×250.
(Reduced to ½ size in reproduction)

CHROMATIC ABERRATIONS

Even though all the spherical aberrations have been well corrected, a lens might yet be at fault because these corrections had been made for one particular wavelength of light only. The focal points for various wavelengths of light might be separated along the optic axis (longitudinal chromatic aberration) or the magnifications achieved with the various wavelengths might be unequal (lateral chromatic aberration). In either case, the image would be blurred into a series of overlapping coloured images. The degrees of correction applied in practice have been described in the body of this chapter, because it is this correction that gives the names of the various types of objectives.

APLANATIC LENSES

An aplanatic lens is one that is corrected for spherical aberration (over the whole of the field) and completely for coma over two-thirds of its diameter, the remaining third also being very nearly free from coma.

4 Practical Adjustment of the Microscope

In the previous chapter the principal components of the metallurgical microscope have been described and brought together in such a way as to give some idea of the theoretical basis for their operation. In the present chapter it is intended to show how the theoretical ideas are made to work in practice—to lay down some guiding rules for the "tuning" of a microscope for optimum performance.

Modern microscopes are becoming increasingly compact and automated, so that the operator has less control over the adjustment of illuminating conditions and other factors which affect the quality of the final image. In buying a microscope, therefore, it is well to know what degree of flexibility is being traded for stability, robustness, compactness or automatic operation. This knowledge can perhaps best be gained by appreciation of the adjustments which can be made on an "optical bench" type of microscope such as we have used as the prototype in the previous chapter.

Illumination

The attributes of the various kinds of sources of illumination have already been discussed; all are satisfactory from the point of view that they will fulfil the theoretical requirements for good illumination. The choice will be made—if a choice is offered—on the basis of cost, intensity of illumination or reliability.

The next requirement for good illumination is centring the source with respect to the axis through the condensers, stops and lenses. It should, of course, be clear that the mechanical arrangement of a good microscope will be such that no such adjustment is needed to align the objectives and oculars. To centre the illumination a specimen is put in focus and the field stop closed until its shadow can be seen superimposed on the image; it should then appear centrally disposed. Many microscopes have an adjustment, by means of set-screws, which can centre the field stop if this is necessary. Once this adjustment has been made there should be no reason for having to re-adjust subsequently. The image of the field stop should be sharp; if it is not, it should be moved along the optic axis until it is sharp, but the great majority of microscopes do not have this movement.

The diaphragm of the field stop can now be used, when closed, as a screen to show an image of the lamp (or the condenser C_1); the lamp housing can be moved to tilt the source until its image is centrally placed on the diaphragm. Final adjustments can be made after the attainment of critical illumination, described below, or when the image is being focused on the screen for photography. It is best to rock the source to and fro by means of its adjusting screws so that the position finally chosen lies symmetrically between the regions in which illumination falls off. The most searching test for even illumination is a photograph of a specimen, but it is as well to get as near to even illumination as possible before taking a photograph.

CRITICAL ILLUMINATION

Originally critical illumination meant the attainment of phase-agreement (coherency) of the light, but this usage is now outmoded and the term is used to describe the condition for full and even illumination of the part of the specimen viewed. This is reached by ensuring that the rear component of the objective is filled with intense and evenly distributed light. Not many microscopes allow any adjustments to their illuminating systems, and they are therefore constructed to have a form of critical illumination built in. With the optical-benchtype of microscope it is achieved (Fig. 20) by moving the condenser C_1 to image the source on

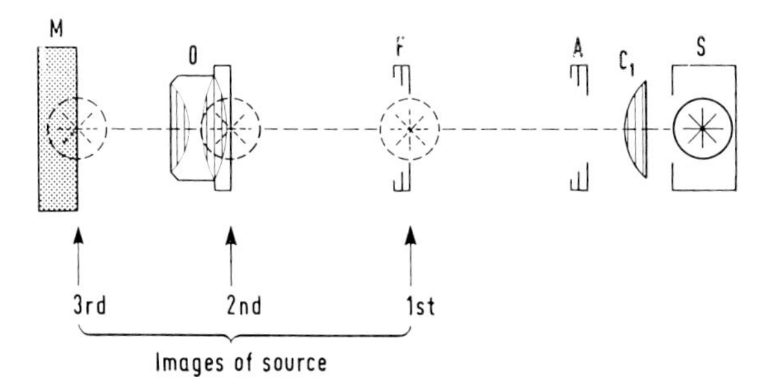

Fig. 20. Critical illumination—the source S is imaged on to the field stop F, fills the rear focal plane of the objective O with uniform illumination and consequently gives even illumination of the portion of the specimen viewed.

the field stop. This then leads to a second image at the next conjugate point, at the rear of the objective, and then to a final image on the specimen itself when it is in focus. Critical illumination achieved in this way requires a source sufficiently large and uniform to produce an image which fills the rear aperture of the objective. Clearly, any self-structure of the source, such as filament coils, or bubbles in the transparent bulb of an arc, will be superimposed on the final image of the specimen. This defect is overcome in another form of critical illumination, which will now be described.

KÖHLER ILLUMINATION

This alternative method images the condenser C_1 on to the field stop and thus on to the specimen. In effect, the condenser is made the source (see Fig. 21); for this to be so it must be of relatively short focus and be placed close to the source. Köhler illumination is very commonly employed in metallurgical microscopes.

INTERMITTENT UNEVEN ILLUMINATION

There are many causes of the uneven illumination which may arise from time to time. Obvious ones are connected with faults in the lamp—sagging filaments, wandering arcs, etc. Water-filled heat absorbers may give trouble, either from evaporation when the meniscus can cross the

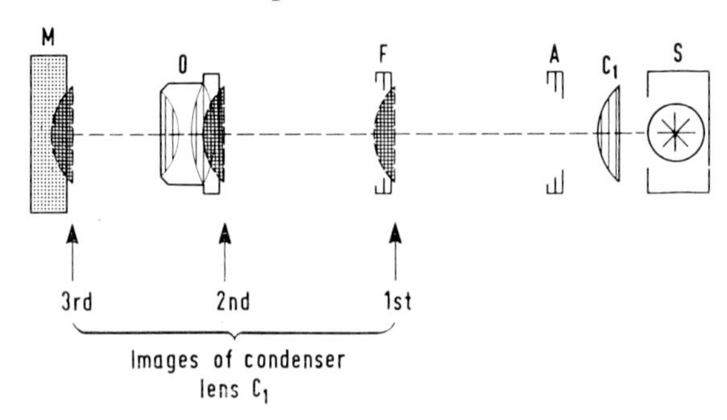

Fig. 21. Köhler (critical) illumination—the condenser C_1 is imaged on to the field stop F and thus achieves even illumination as in Fig. 20 but with the condenser C_1 replacing the actual source.

beam or from growth of algae or from precipitation of salts on the glass walls of the trough. With properly constituted liquid in the filters such things should not happen, but if they do it is not always obvious where the trouble lies. Distilled water in which a small crystal of copper sulphate has been dissolved is recommended to control these defects.

Excessive dirt, or scratches, on glass surfaces which are close to conjugate points may also cause uneven illumination, but it does have to be a major obstruction before deterioration is obvious in the image.

Tube Length

Clearly, the focal lengths of objectives and oculars are such that they have to be separated by a specific distance in order that a virtual image is produced at the eye or a real image on the viewing screen. The requisite optical distance is known as the optical tube length, which is defined as the distance between the rear focal plane of the objective and the front focal plane of the ocular. In the earlier transmitted-light microscopes, objectives were paired with oculars so that the image of the specimen was made to appear to the eye as if it occupied the actual position of the specimen, and thus the optical tube length was made 250 mm, the distance for the nearest distinct vision. Subsequently, microscopes were made more compact and an optical tube length closer to 150 mm became more common. The attainment of the correct optical tube length is reached through use of an appropriate mechanical tube length. This is defined as the distance between the shoulder of the objective, where it rests on the holding ring, and the top of the draw tube, where the end of the ocular rests.

Use of the wrong tube length can impair the performance of a microscope quite markedly, but it is rare for any problem to arise, because each make of microscope is constructed with a fixed tube length to match the range of objectives available in that brand. However, some microscopes do have movable draw tubes and their setting must be watched. Trouble can arise, moreover, if objectives from one microscope are used in another; in particular, objectives from transmitted-light microscopes are usually made to be used with a tube length of 160 mm and are therefore unsuitable for use on a metallurgical microscope. Even so it is only in the more critical applications that deterioration would be noticed. The tube length is often marked on the objective.

Several well-known makes of microscopes use objectives designed to work with an infinite tube length. This is presumably used to give extra room for illuminators, analysers, etc. and may have advantages in avoiding distortion at such components. There is also an advantage in connection with the design of associated equipment for phase contrast. It means, however, that interchangeability with other makes of objectives is restricted for these microscopes.

Calibration of Magnification

The magnification given by a compound microscope can be found by multiplying the magnifications of its components, the objective and ocular. Most lenses have their magnifications marked on them, but it must be remembered that the magnification for the objective refers to a particular tube length.

It can be shown very simply (see, e.g., Martin and Johnson[1]) that the magnification M is given by

$$M = tD/fe \qquad (1)$$

where t is the optical tube length, D the distance of distinct vision for the unaided eye, and f and e the focal lengths of the objective and ocular respectively. The magnifying power of the objective is t/f and of the ocular D/e, so this formula is merely a restatement of what has been said above. Unfortunately, some makes of lenses bear D/f for the magnification of the objective and t/e for the ocular, so the magnification can be confusing. The marked magnifications are intended only as a guide and should never be used except as such. It is best, therefore, to measure magnification directly; this is particularly desirable when a bellows extension is involved as well. To do this a "stage micrometer" is used as a specimen. It consists of a scale engraved or photographically reproduced on a strip of metal or glass. The image of part of the scale can then be measured on the focusing screen or on a second scale in a suitable ocular (a "filar eyepiece"). It is convenient to plot magnifications against bellows lengths for various combinations of objectives and oculars, but this is perhaps more useful as an averaging procedure to smooth experimental errors than as a guide to finding magnifications. This remark is made to introduce the thought that standardization of magnifications in metallography is very desirable; an A.S.T.M. recommendation gives the following standard magnifications: 25, 50, 75, 100, 150, 200, 250, 500, 750, 1000, 1500 and 2000. There seems to be no reason why these values should not be adopted for the great majority of micrographs.

Selection of Magnification

The achievement of critical or Köhler illumination enables the best use to be made of a given objective provided that it is properly matched to its ocular. However, some thought should be given to the selection of the best objective for a particular specimen. It has already been remarked that resolution is not often the critical factor in metallography. When it is so, all the skill and knowledge of the microscopist is required to gain the last fraction of a micron from the equipment available.

With the best lenses (apochromats) and very great care both in preparation of the specimen and in adjustment of the microscope, it is possible to work with magnifications up to 1500 (N.A.)—that is, with total magnifications of $\times$ 1500 to $\times$ 2000.

We can estimate the minimum magnification (M_m) required to allow the eye to perceive the resolution achieved if we know the resolution the eye can itself achieve. This is often taken to be about $1{\cdot}1 \times 10^{-2}$ cm. Suppose the microscope objective has an N.A. of 1·0 and the light used a wavelength of $5{\cdot}4 \times 10^{-5}$ cm. Equation (8) of Chapter 2 tells us that the resolution for these conditions is $3{\cdot}3 \times 10^{-5}$ cm. Therefore, the magnification M_m which would enable the eye to resolve the image of two points which are $3{\cdot}3 \times 10^{-5}$ cm apart is $M_m = 1{\cdot}1 \times 10^{-2}/3{\cdot}3 \times 10^{-5} = 330$. Since N.A. = 1, we can write this as 330 $\times$ (N.A.), so that this example suggests that the minimum magnification required for the eye to appreciate the resolution achieved by the objective is 330 times the objective's N.A. This may be shown to be a general result, using an argument outlined in the next paragraph, although assumptions about the geometry of the eye may lead to values of M_m as low as 250 $\times$ (N.A.).

The argument is based on the observation that the nearest distance for distinct vision is about 250 mm from the eye and the diameter of the pupil of the eye between 1·5 and 2 mm. With a diameter of 1·5 mm the N.A. of the eye is approximately 0·75/250, because here $n \sin \theta \simeq n \tan \theta$ (θ being small) and $n = 1$. Use of Lagrange's Equation (see p. 14) leads to (N.A.)/(0·75/250) $\simeq$ 330 $\times$ (N.A.). Taking the diameter of the pupil to be 2 mm gives $M_m \simeq 250 \times$ (N.A.).

In practice this means that, since the magnification of the objective is fixed, additional magnification to bring the total to 330 $\times$ (N.A.) has to be provided by choosing an appropriate ocular and the necessary

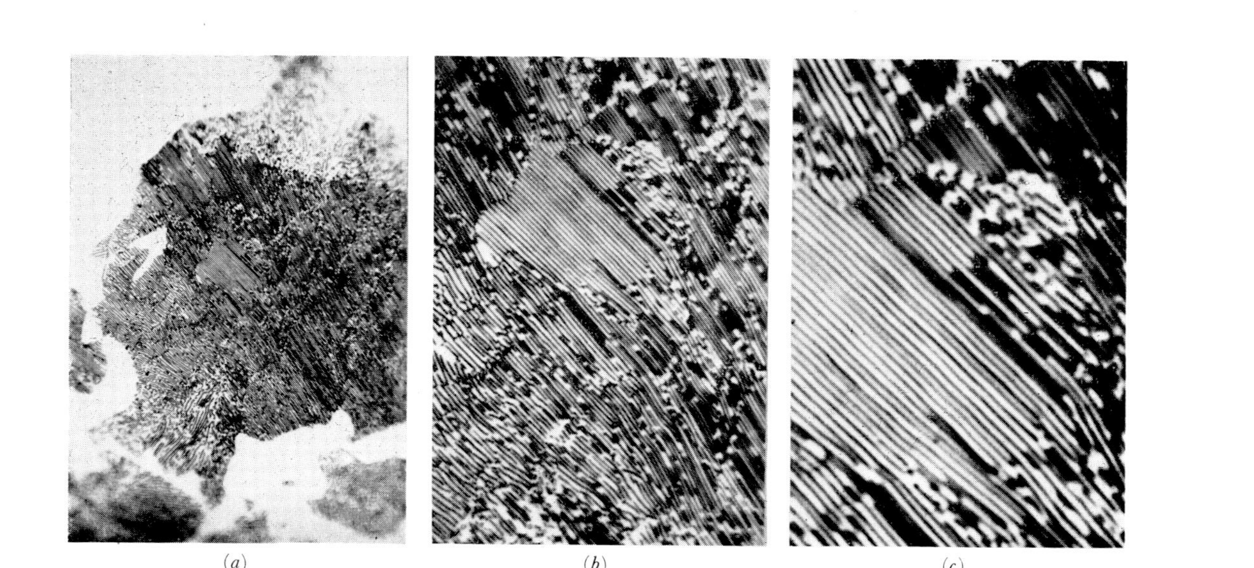

Fig. 22. Total useful magnification as a function of N.A. (here 1·25) illustrated using an etched pearlitic steel.

(*a*) $\times$500, equal to 400$\times$ (N.A.).
(*b*) $\times$1000, equal to 800$\times$ (N.A.).
(*c*) $\times$3000, equal to 2400$\times$ (N.A.).

Note that in (*b*) it is easier to perceive detail than in (*a*) although the resolution is unchanged; moreover, in (*c*) there is no improvement in clarity, only empty magnification, which has an advantage only if the print is viewed at distances greater than $\sim$250 mm (10 in.).

(Reduced to $\frac{9}{10}$ in reproduction)

bellows extension. If a bellows extension is not possible, the final stage of magnification may be enlargement of the negative during printing of the photograph. This should then give a separation of closest points of 0·5 mm if the print is to be viewed at a distance of 250 mm. Usually, a magnification of two or three times the theoretical minimum is tolerable or even advantageous to those with normal eyesight. This does not, of course, improve resolution, but merely makes it easier for the eye to sort out the detail: it is *empty magnification.* The empty magnification should be increased still further if the print is to be viewed at distances greater than 250 mm, as would be the case for prints in an exhibition.

For much metallographic work, when the emphasis is upon the general inter-relationships of structure rather than on the ultimate detail, a magnification of 500 × (N.A.) would seem to be acceptable. This requires reasonable care in adjustment of the microscope, but not the extreme precautions demanded by the use of 1500 × (N.A.). Electron microscopes of modest power and cost give magnifications of × 2000 and, moreover, with much improved resolution compared to the optical microscope, so that it may not seem worth while attempting to push the resolution in the visible spectrum to the very limit. The effect of magnification in relation to N.A. upon resolution is illustrated in Fig. 22.

Samuels[2] has condensed the foregoing discussion into a set of working rules, which are summarized here.

(1) Magnifications up to × 500—select an objective which will operate at a magnification in the range 400–750 times the N.A. The lower the N.A. of the objective, the more desirable it is to operate at the lower end of the range.
(2) Between × 500 and × 1500—select an objective which will operate at a magnification in the range 750–1200 times the N.A. The higher the N.A. the more desirable it is to operate at the high end of the range.
(3) For magnifications up to × 2000—use only the highest quality apochromats (N.A. 1·3–1·4) and select the most critically adjusted conditions.

Achromatic lenses should not be used at magnifications above × 750.

Choice of Oculars

To a great extent the choice of the correct ocular is dictated by the choice of the objective which has been made, since only specific combinations of the two kinds of lens give full correction in many cases. If the microscope has an adaptor to vary the tube length, it is important to see that this is correctly set for the lens used.

At low magnifications, when achromats are used, uncorrected oculars may be used. It is best to limit the eyepiece magnifications to × 10 or × 12·5, with a maximum of × 15. In any case the overall magnification should not be forced above × 500 when using achromats of low power.

With the higher-powered achromats, intermediate oculars should be used. For the high-powered apochromats it is necessary to use the appropriately corrected compensating eyepieces. It is worth testing the various combinations for signs of distortion, such as uneven magnification or lack of flatness of field.

Filters

It is seldom that any filter other than a yellow-green one is used in metallography. The wavelength of green light is squarely in the middle of the fully corrected range for all types of objectives and the eye has its maximum acuity in the yellow-green light. It is also within the range of maximum sensitivity of the most widely used types of photographic emulsions. Furthermore, the types of illumination commonly used have good intensity in the green region of the spectrum; it is the reason, at least in part, that they are commonly used.

There are occasions when the small but positive gain in resolution makes it worth using a blue filter, but focusing can then become very difficult because of the lower intensity on the screen. The focusing magnifier (q.v.) should be used in such cases. If a blue filter is used it must be remembered that a photographic emulsion with suitable characteristics is required to record the image.

Use of Field Stop

The sole purpose of the field stop is to reduce flare or unwanted reflections by limiting the illuminated area of the specimen. The stop should be opened until its image is just outside the field of view.

Use of Aperture Stop

Unlike the field stop, the aperture stop can have a marked effect on resolution or may even introduce spurious detail into the image. Once critical illumination has been achieved with a particular illuminant the only way intensity can be further controlled is through closing the aperture stop. However, this also reduces the disc of illumination at the rear of the objective lens, and so reduces the N.A. of the system. In this way resolution is reduced. If such a reduction can be tolerated, the lower intensity may assist in increasing contrast because less scattering of light occurs. It may also increase both the depth and flatness of field.

Microscopists seem to be in agreement that a slight closure of the aperture stop is worthwhile; 90 per cent of the diameter of the back lens of the objective is often quoted as the optimum size for the illumination at that point. With modern optical equipment, for which "blooming" of glass surfaces to reduce unwanted reflections is usual, it is doubtful whether this slight reduction in aperture does, in fact, have any noticeable effect on residual flare or scattering.

If the aperture stop is closed down to very small sizes, marked diffraction fringes are formed at many features. Inexperienced metallographers often mistake these fringes for sharpening of detail (e.g. of carbide particles in steels). Closing the stop to this degree is bad practice, although it can be made to yield useful information if the microscopist understands what he is doing. This point will be taken up again in the next chapter. A skilled microscopist may make small adjustments to the aperture stop as he scans a field, but these are used to aid him in picking out detail rather than in recording it. The setting corresponding to 90 per cent or 100 per cent of the rear lens has to be found for each objective. This can be checked by removing the eyepiece and viewing the image of the stop which appears on the rear lens. These aperture settings should be recorded, for they are essential if standardized photographic exposures are to be used (see Chapter 8).

Focus

It would seem to be somewhat trivial to remark that part of the critical adjustment of the microscope is correct focusing. But focusing is not a straightforward operation, particularly for photography. It depends on acquired skill in balancing the defects in the image to produce a tolerable record, and although it should be sufficient merely to record the desired feature in isolation, it is nevertheless true that the skilled metallographer increases the impact and acceptability of his record by relating it to the surrounding detail and by giving it some aesthetic appeal. When viewing a specimen by eye, continual changes in focus and even in lighting conditions can assist in revealing meaningful detail. The photographic record then becomes equivalent to a "still" from a cinema film and may fail as a record unless additional skill is employed in making it. For photography the image should be roughly focused on to a ground-glass screen and the fine focus used to move to and fro through true focus, obtaining the best overall realization of the desired detail. This should be done with the appropriate filter in position, even though this reduces the intensity of illumination.

When the maximum resolution of detail is required, focus should be made at the centre of the field and the inevitable curvature of the field accepted. It sometimes helps to resolve detail on the screen if it is gently oscillated in its guide to blur the structure of the ground glass; however, final focusing should be carried out subsequently, so that no risk of disturbance by vibration is incurred. For the greatest precision in focusing, the ground-glass screen can be replaced by a piece of plane glass and the image examined by a special focusing magnifier. This is placed on the glass and usually, but unnecessarily, focused on the surface of the glass corresponding to the plane which the photographic emulsion will occupy. The point is that the projected image of the specimen is not in focus only at the plane of the photographic emulsion; it is, in fact, in focus when viewed on a screen placed anywhere beyond the projection eyepiece, provided it has been focused for one such position. The magnifier is usually focused by rotation of its eyepiece component. The image of the specimen is then brought to sharp focus when viewed by the magnifier. Since this procedure is seldom used unless the highest magnifications are being employed, it may often result in a relatively small area appearing in focus, although modern metallographic equipment is designed to give acceptable sharp images over the area of at least a quarter-plate negative (4 in. × 3 in. approx.). If the focusing technique just described yields a much smaller field than this in sharp focus, some faults in technique or equipment must be suspected.

A common fault is curved or tilted specimens. A high-power objective cannot accommodate any but the smallest changes in level of this kind (see "Depth of Field", below).

It is good practice to focus roughly by bringing the specimen slightly closer to the objective than necessary for true focus, using the coarse focus and observing this directly by eye. Many microscopes have marks to indicate this position. Then, whilst viewing the image through the eyepiece, true focus is approached using the fine focus to move the specimen *away* from the objective. This reduces the risk of damaging the objective by jamming it against the specimen. Thus, one of the first lessons in microscopy is to learn which way to turn the fine focus in order to move the objective away from the specimen; after a head-on collision, it is little consolation to know that it is easier to repolish a scratched specimen than a damaged lens. If there is noticeable backlash in the fine focus screw, final focusing should be made against this backlash.

Depth of Field

It is possible for an objective to bring various levels of a region of a specimen into acceptable focus simultaneously if certain geometrical requirements are met. As might be expected, these relate to the wavelength of light and the N.A. Suppose the microscope to be focused on the plane of an object which is normal to the optic axis and passes through S_1 (Fig. 23). Now let the attention be shifted to another plane through S_2. Rayleigh suggested that the paths of the marginal rays in the image space OI should differ by no more than $\lambda/4$ for sharp focus to be retained. A greater difference would lead to a broadening of the bundle of rays

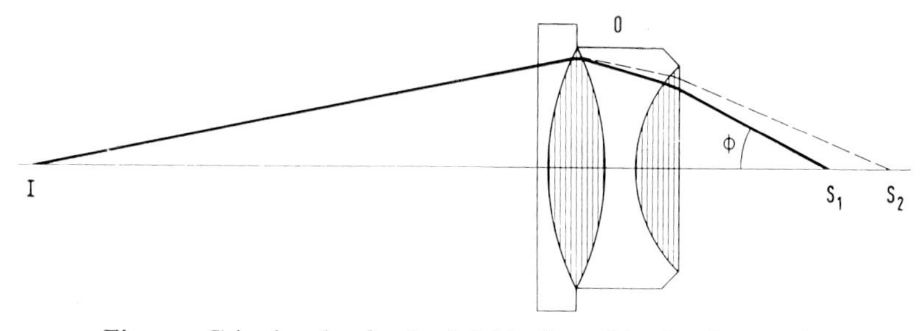

Fig. 23. Criterion for depth of field of an objective (see text).

which should give an image of a point larger than the circle of confusion introduced by diffraction. On the basis of this criterion it can be shown that the corresponding depth of field $S_1S_2 = d$ is given by

$$d = \pm (\lambda/4)/(n \sin^2\phi) \approx \pm n\lambda/(\text{N.A.})^2 \qquad (2)$$

where $\lambda/4$ is the allowable path difference defining *depth of focus*, n the refractive index of the object space OS_1 and ϕ the angle OS_1I.*

It is found in practice that a depth of field based on the path difference $\lambda/4$ is well within the range for acceptable focus; somewhat greater differences can usually be present without noticeable deterioration of sharpness of focus.

Martin and Johnson[1] have computed the following table to show how depth of field depends upon N.A. for a particular wavelength of light ($\lambda = 0{\cdot}5$ μm).

TABLE 1

Effect of N.A. on Depth of Field

N.A.	Depth of Field in μm	
	In air	In oil ($n = 1{\cdot}5$)
0·25	7·9	12·2
0·50	1·9	3·0
0·75	0·8	1·3
1·00	—	0·7
1·25	—	0·4

The limited depth of field, especially of objectives of high N.A., is sometimes a little troublesome in that it necessitates continual refocusing as a specimen is scanned, but presents fewer problems when a specimen has been cleanly polished and lightly etched. Sometimes, as on fracture surfaces, interest does centre on a range of levels. Visual examination is possible, using changes of focus, but normal photography does not give very informative results. Alternative ways of recording such structures are the use of the stereomicroscope or of plastic replicas and electron microscopy. It is also possible to examine small pieces of such replicas; by this means the short working distance of the optical microscope and its limited depth of field might be accommodated.

If an accurate assessment of the depth of field is required, some specially designed test object must be used. Shillaber[3] recommends a mark drawn

* Shillaber[3] quotes alternative formulae. The first for photographic work is

$$d = [(n^2 - (\text{N.A.})^2)^{\frac{1}{2}}]/(\text{N.A.})^2,$$

and the second for visual work, where the accommodation of the eye is such that a greater depth of focus can be tolerated, is

$$d = 250 \text{ mm}/M^2$$

where M is the magnification produced by the microscope.

along the line of greatest slope of a right-angle prism, with the sloping face about 30° to the horizontal. With metals, it is possible to make use of diamond pyramid indentations, calculating their depths from the known geometry of the 136° pyramid and the measured diagonals of its impression on the metal.

Care of Optical Equipment

In discussing focusing we mentioned an elementary aspect of the care of optical equipment. There are other ways in which expensive optical parts may be damaged. Dropping them is one. Lenses may have their efficiency impaired by dust, but a more usual cause of poor images is finger prints. The microscopist must learn to handle lenses without fingering the glass surfaces. This would seem a simple matter, but it is surprising how often finger prints do appear on objective lenses, let alone eyepieces. Such would, no doubt, be most valuable to anyone investigating murder in the microscope room, but this is a small advantage to set against perpetually degraded images. Dust can be removed with a puff of air from a rubber bulb and grease with a solvent such as xylene, benzene or dimethyl ketone; (do not use alcohol, as this may attack the cement between lens components). With other solvents, use as little as possible, for these also attack the cement in time. Xylene and benzene are toxic, especially so to some people; there is no risk from sparing and infrequent use, but it is best to ban them from the laboratory and use dimethyl ketone.

To remove grease or immersion oil, place a lens tissue over the lens and drop a little of the solvent on to it, then draw the tissue slowly across the lens. Repeat this with fresh pieces of dry tissue until the glass is dry.

Perhaps it is worth mentioning the correct way of using oil for an oil-immersion lens. Such objectives are clearly marked for use with oil and the maker usually supplies or recommends the correct kind of oil. This is usually a cedar oil. It is convenient to have this in a bottle with a glass rod through the stopper, so that a drop of oil can be allowed to run off the rod on to the front component of the objective. A second drop is allowed to drop on to the area of the specimen to be examined and then the objective moved towards the specimen until the two oil films merge; this is judged by eye. Final focus is then achieved in the normal way, moving the fine focus very slowly, for the range of recognizable focus is small and can be easily overshot.

Dust or blemishes on other parts of the optical system may give a more obvious effect on the image than they do when on the objective. Particularly vulnerable parts are those close to conjugate points—i.e. those points which are brought to a focus on the specimen or where images of the specimen are formed. These include the condenser C_1, the vertical illuminator and parts of the projection eyepiece. The dust particles will probably appear out of focus as blurred discs or hazy worms on the image, and may not be noticed until negatives are examined; the same tell-tale shapes will then appear on each of a sequence of photomicrographs. Although this may not, in fact, confuse the record of the structure, it is indicative of sloppy technique and should be corrected.

Dust particles of somewhat larger size may cause trouble in focusing to obtain a flat field, by tilting the annulus on which the specimen rests. It should also be remembered that the fine focusing mechanism is making adjustments which are comparable with the wavelength of light; it has, therefore, to be treated as a precision piece of engineering and kept clean, free from coagulated grease and not subjected to severe stresses.

REFERENCES

1 L. C. Martin and B. K. Johnson, *Practical Microscopy*, London, 1958 (3rd edn) (Blackie).
2 L. E. Samuels, *Australasian Engineer*, 1949 (Nov.), **42,** 134.
3 C. S. Shillaber, *Photomicrography in Theory and Practice*, New York, 1944 (Wiley).

5 Bright-field and Related Techniques

The layout of the microscope which has been described in the preceding chapters is that for normal or bright-field illumination, the word normal, in fact, referring to the angle at which the beam meets the specimen but, perhaps, being equally acceptable in its sense of meaning usual. For the greater part, the average metallographer will find bright-field illumination is all that is required, most specimens being essentially two-dimensional and having features differentiated by reflectivity, structural texture or colour. Further, except when discussing resolution, geometrical optics suffices to explain the image formation in such cases. The other techniques to be described in this chapter form a family, the relation between its members being brought out in Fig. 24; the father of this family is the Abbe theory of the microscope.

It will be seen in the sections which follow that the major group of illuminating techniques depends upon forming the image from various combinations of the direct and diffracted beams from the object. This group contains types of bright-field as well as dark-field and oblique illuminations. Because of their dependence upon the combination of direct and diffracted beams in various proportions, they may be spoken of as *interference-balance techniques*. In these it is the amplitudes of the various beams which are being controlled; in contradistinction, the other group is characterized by control of the phases of the direct and diffracted beams. In consequence, this group comprises *phase-contrast techniques*. We will first consider a series of techniques related to bright-field.

"Out-of-focus"

In discussing the aperture stop (p. 41) it was pointed out that a spurious increase in contrast can be obtained by closing this stop down. What happens is that diffraction patterns become visible. The reason differs according to the kind of specimen. For very small features, like brightly reflecting constituents or pits, the ever-present diffracted beams become visible as the intensity of the general illumination is decreased and so diffraction maxima and minima are seen surrounding or alongside these features (Fig. 25).

Similar but somewhat broader fringes are also seen under these conditions at small changes of slope on the surface of specimens, such as sub-grain boundaries on specimens which have been polished and then deformed during creep (Fig. 26). The amount of defocusing required is

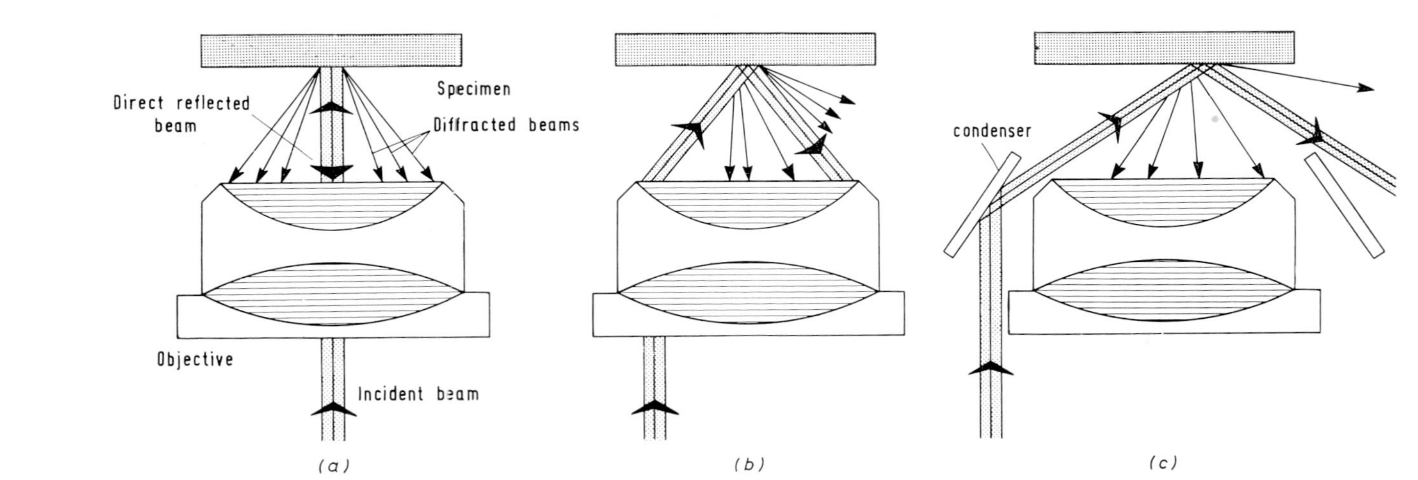

Fig. 24. Schematic representation of "interference-balance" methods of illumination, showing their relation to normal illumination.

(*a*) Normal, bright-field.
(*b*) Oblique.
(*c*) Dark-ground (or dark-field), sometimes called oblique dark-ground, obtained using a catoptric condenser.
(*d*) Sensitive dark-ground (after Lomer and Pratt) using opaque stop.
(*e*) Stop-contrast (after Hallimond).
(*f*) Phase-contrast.

Except in (*a*) the diagrams have been simplified by showing only one half of the beams.

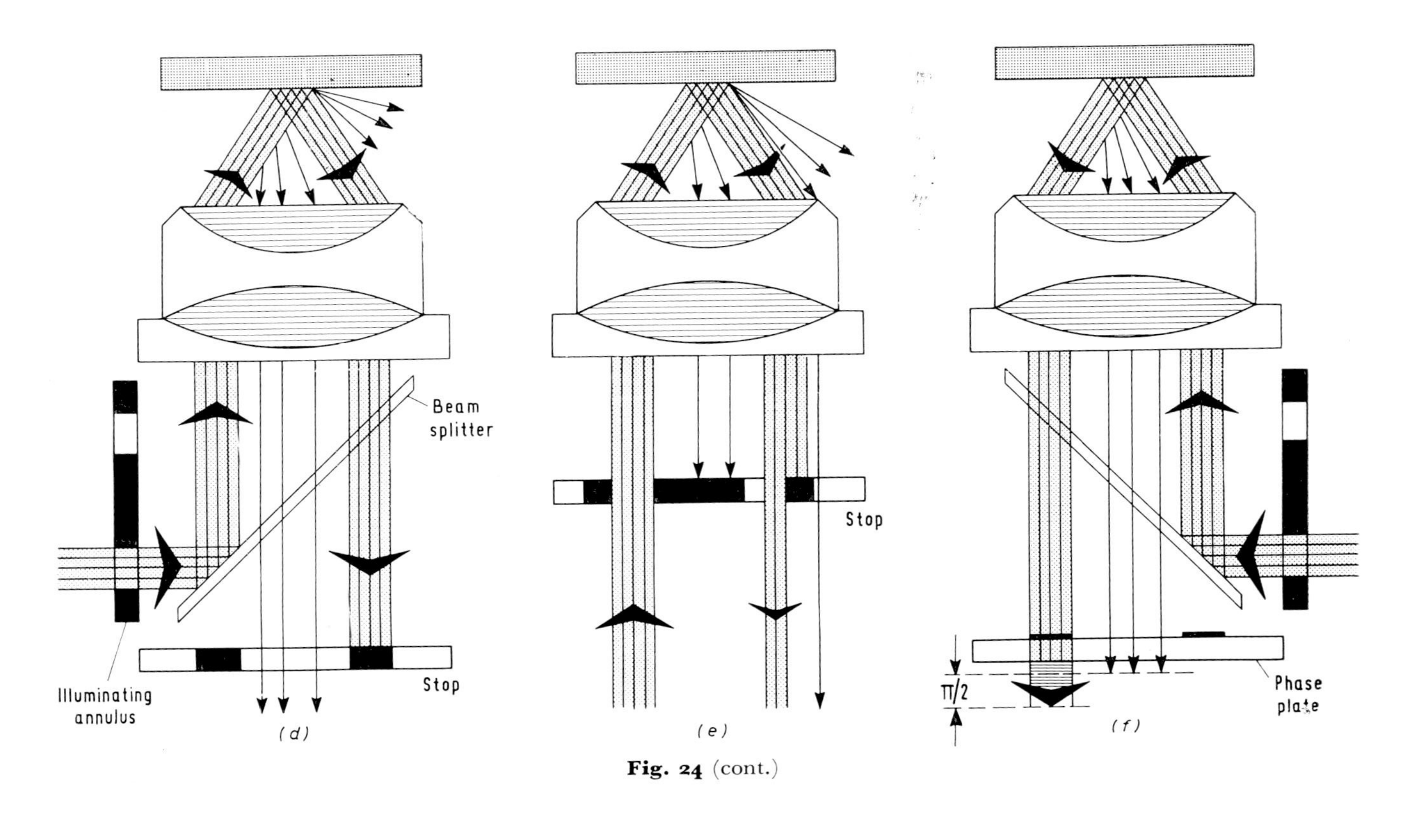
Beam
splitter
Illuminating
annulus
Stop
(d)
Stop
(e)
π/2
Phase
plate
(f)

Fig. 24 (cont.)

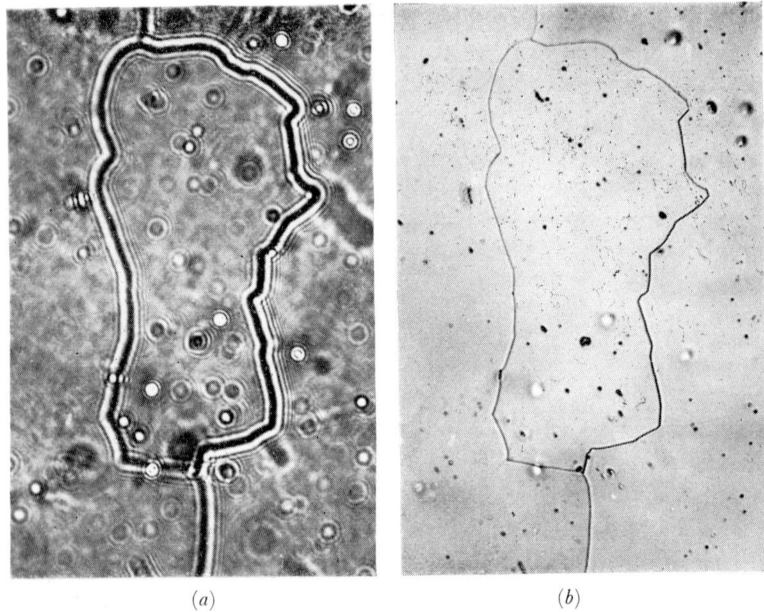

(a) (b)

Fig. 25. (a) Diffraction effects at grain boundaries and at pits of various sizes, revealed by closing the aperture stop.

(b) The same area with the aperture opened to its optimum position.

The specimen is electrolytically polished aluminium; unetched, $\times$ 100.

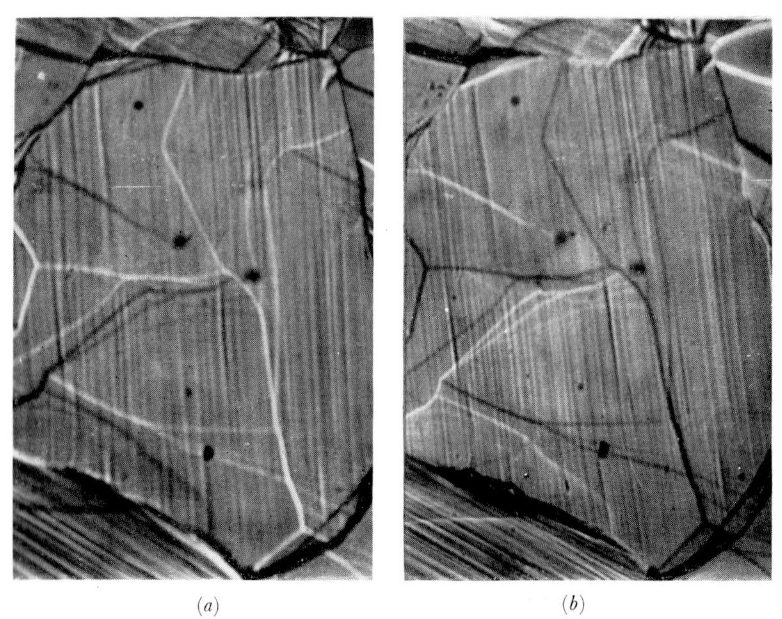

(a) (b)

Fig. 26. "Out-of-focus" technique, aperture stop closed. Small changes in level at slip lines and sub-grain boundaries are revealed as black or white lines. Note how black and white interchange when the focus is changed from one side of true focus in (a) to the other side in (b). The larger steps at grain boundaries remain black in both photographs. The specimen is electropolished zinc deformed during creep; unetched, $\times$ 150.

quite small, of the order of 1 μm, as considerations of depth of field might lead us to suppose. Moreover, the fringe contrast reverses on changing from one side of true focus to the other, as illustrated in Figs. 26 (*a*) and 26 (*b*), black lines becoming white and vice versa. This effect is, of course, closely related to the symmetrical fringes discussed in Chapter 4 in connection with focusing, but they are asymmetrical here because the specimen is not flat. A full discussion of the theory of these fringes has been given by Berek[1]. The angle between the faces giving the effect on specimens such as that used for Fig. 26 is small; it has been measured interferometrically and found to be $\sim 30'$.

This technique is thus very useful for the *detection* of small features such as grain-boundary cavities formed during creep, or of small surface

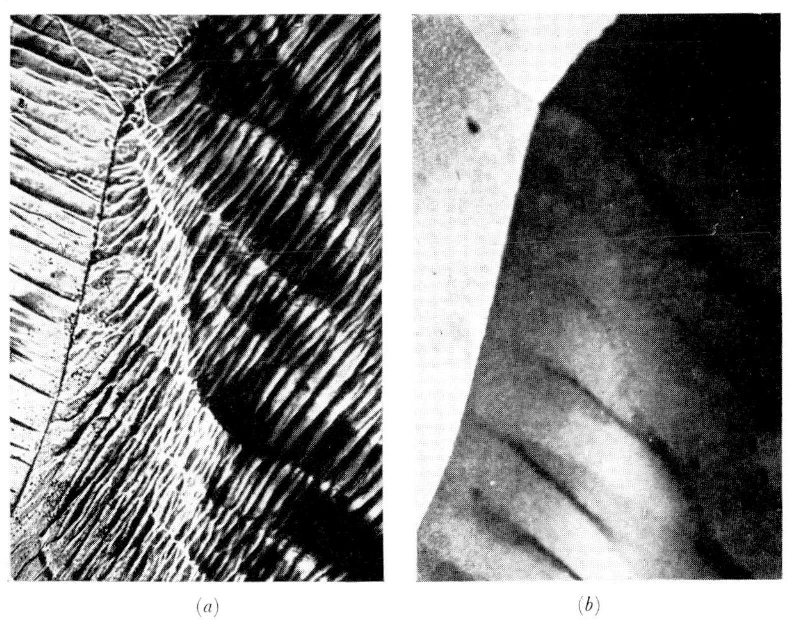

(*a*) (*b*)

Fig. 27. (*a*) "Out-of-focus" technique, aperture stop closed. Here, the deformation markings on an aluminium creep specimen are more complex than those for zinc in Fig. 26, $\times$ 150.

(*b*) Repolishing, anodizing and viewing with polarized light (see Chapter 7) help to sort out the detail in (*a*), by showing the bands of sub-grains, $\times$ 150.

irregularities, but it must be remembered it does no more than detect them and does not give a true image. Moreover, it is so sensitive that it may sometimes give a complex pattern which confuses significant features in a maze of minor ones (Fig. 27 (*a*)). In that case, a "second opinion" must be sought through the use of some other technique (Fig. 27 (*b*)).

Interference-balance Techniques

This is an important group of techniques, some old and well-tried, some new, which depends upon changing the pattern of destructive interference in the image in another way. As mentioned before, this is done by alteration of the balance between the direct, undiffracted, beam and the diffracted beam. Although the complete reconstruction of the object in the image requires that all of these beams be collected and allowed to interfere, it is possible to get some kind of image with any pair of them. These techniques will now be described and their inter-relationship made clearer by the series of diagrams in Fig. 24.

OBLIQUE ILLUMINATION

In Fig. 24 (*a*) bright-field illumination is represented; the incident beam is normal to the specimen and thus the direct beam is reflected normal to the specimen surface. Several diffracted beams, symmetrically arranged about the direct beam, are also collected by the objective lens; their number, of course, depends upon the N.A. of the lens and the wavelength of the light employed.

If the diffracted beams on one side of the direct beam are blocked off in some way, the image shows a shadowing effect. This condition can most readily be obtained by making the incident beam meet the specimen obliquely, as in Fig. 24 (*b*). On some microscopes oblique illumination can be achieved very simply by moving the aperture stop (on an appropriate carrier) across the beam. The amount of decentring may be measured on a vernier scale and the orientation of the oblique (decentred) beam varied by rotation of the carrier. Fig. 28 shows a zinc specimen after creep, with a structure such as shown previously in Fig. 26, but here using oblique illumination obtained by decentring the beam.

A second way of obtaining oblique illumination is by the use of the *patch stop* or *conical stop*. This consists of an opaque disc inserted into the illuminating beam close to the aperture stop. It is therefore imaged close to the plane of the rear lens of the objective and the objective then focuses a hollow cone of rays on the specimen, as distinct from the solid cone in bright-field illumination. The effect is thus the same as that illustrated in Fig. 24 (*b*) but the obliquity is now symmetrical. It has the possible disadvantage of failing to use the full angular aperture of the objective, although this is not often very important; this is because the great majority of applications for oblique illumination seem to be associated with total magnifications of × 300 or less, and thus the loss in resolution is not likely to be noticed. If the stop is moved away from its central position, the illumination will become more oblique and may then be almost exactly the same as that illustrated in Fig. 28.

A third, but possibly less convenient, way of obtaining oblique illumination would be that of tilting the specimen. Some bench microscopes have a special tilting stage mounted on a slide which fits on the normal stage; this is primarily for use with phase-contrast illumination, but it could be used to obtain oblique illumination.

Prism or sector illuminators also give oblique illumination, although the degree of obliquity is small and fixed (see Fig. 29 (*c*)).

Oblique illumination is extremely useful in many applications where

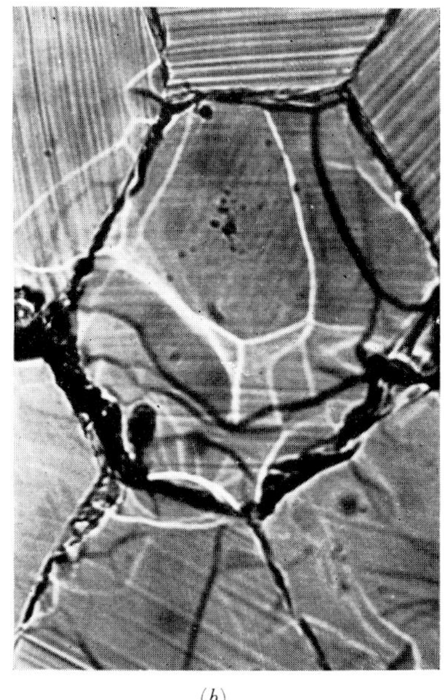

(*a*) (*b*)

Fig. 28. (*a*) Oblique illumination, obtained by decentring the prism illuminator; zinc creep specimen, unetched, × 150.
(*b*) The same field as in (*a*), out-of-focus technique.
For other examples of oblique illumination see Fig. 29, and also Fig. 56 of Chapter 6.

some surface relief is present, particularly therefore with deformed specimens. It appears to be almost as sensitive as phase contrast in revealing fine slip steps and similar features, but does suffer in comparison because of its directionality. It will be clear that a step must lie across the "shadowing" direction in order to be revealed. This is illustrated in Fig. 29, where slip lines on a deformed specimen are shown in (*a*) and (*b*) with two directions of oblique illumination and compared in (*d*) with the same area photographed using a type of phase-contrast illumination.

DARK-FIELD ILLUMINATION

The shadowing contrast which is characteristic of oblique illumination can be made more and more intense by increasing the obliquity of the incident light until the direct beam, as well as the diffracted beams on one side of it, is reflected outside the objective. This then gives a dark-field or dark-ground image of a kind which is best described as "oblique dark-ground". The schematic trace of the rays in Fig. 24 (*c*) will make the reason for this name clear. It is also sometimes called Spierer illumination.

Dark-field illumination is characterized by the production of an image in which contrast is reversed with respect to the bright-field one. As with oblique illumination, some aspects of the image can be predicted from geometrical optics, but it is as well to remember that the essential element

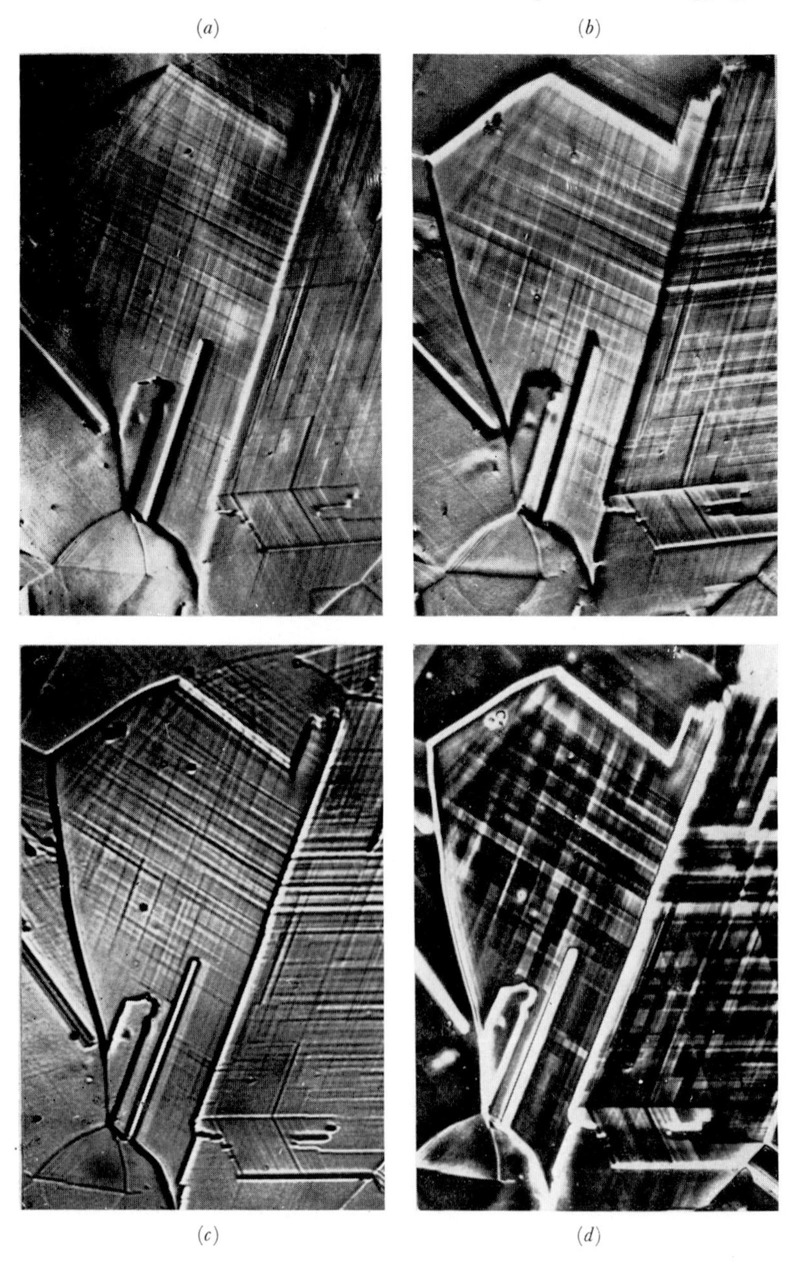

Fig. 29. Oblique illumination compared with phase-contrast, using a specimen of α-brass which has been electropolished and slightly strained. $\times$ 200.

(*a*) Oblique illumination, beam decentred "vertically".
(*b*) Oblique illumination, beam decentred "horizontally".
(*c*) Prism illumination, aperture stop partially closed.
(*d*) Positive phase-contrast.

Note: The same field is shown using other techniques in Fig. 63 of Chapter 6 and a similar field is shown in Fig. 53 of the same chapter.

is the removal of the direct light. Major irregularities, such as pits or scratches, might appear bright because the surface is locally tilted to such a degree that the direct light is brought back into the lens; other areas will have images formed entirely by interference between those diffracted rays which the objective collects.

Because so much of the light is rejected the integrated intensity is low, but this may be an advantage as it reduces scattered light and gives a crisper contrast to the image. For a similar reason it can display the colour of a non-metallic inclusion or precipitated phase rather more clearly than with bright-field illumination. As mentioned on p. 18, the perception of a bright spot or streak on a dark background depends solely upon its intensity, so dark-ground illumination may reveal pits, particles or precipitates more clearly than bright-field.

Many microscopes, as a standard fitting, carry equipment for the production of oblique dark-ground illumination. If often consists of an annular illuminating stop, the diameter of the clear annulus being large enough to produce a hollow cone of rays outside the objective. These are then reflected by an inclined ring of mirror on to the specimen at an angle sufficiently oblique that only one side of the series of diffracted rays is collected by the objective after reflection at the flat specimen. The arrangement is illustrated schematically in Fig. 24 (*c*); it is often known as a catoptric condenser. Other microscopes use double reflections in a glass assembly mounted around the objective to obtain the same oblique cone of rays as in the catoptric condenser.

Despite having been a standard fitting on large microscopes for decades, dark-ground illumination has not often been used, if one may judge from published micrographs. Fig. 30 (*a*) shows an example of the kind of micrograph produced using oblique dark-ground; Fig. 30 (*b*) shows the same area using bright-field illumination.

In the *opaque-stop microscope* Lomer and Pratt[2] developed a means of changing systematically from oblique dark-ground to "sensitive dark-ground", in which symmetry is achieved with only the direct beam removed (see Fig. 24 (*d*)), but in which it is also possible to allow varying proportions of the direct light to be removed, thus producing a continuous range of oblique and dark-ground images. When only part of the direct light is removed this is sometimes called the Schlieren arrangement.

For the normal range of magnifications used on the microscope (i.e. above $\times$ 50) the arrangement used by Lomer and Pratt is illustrated in Fig. 31. The specimen is illuminated with a hollow cone of rays obtained by inserting a stop with a clear annulus near the aperture stop at IA. The annulus is of such a size that the full numerical aperture of the objective lens is used, approximately a third of the area of the back of the lens being shadowed at its centre by the opaque stop. A second stop is incorporated in the microscope at a position conjugate to that of the illuminating stop, but such that only the rays reflected from the specimen are intercepted by it—e.g. at OS in Fig. 31. This stop consists of an opaque annulus of such a size that it would intercept all the direct

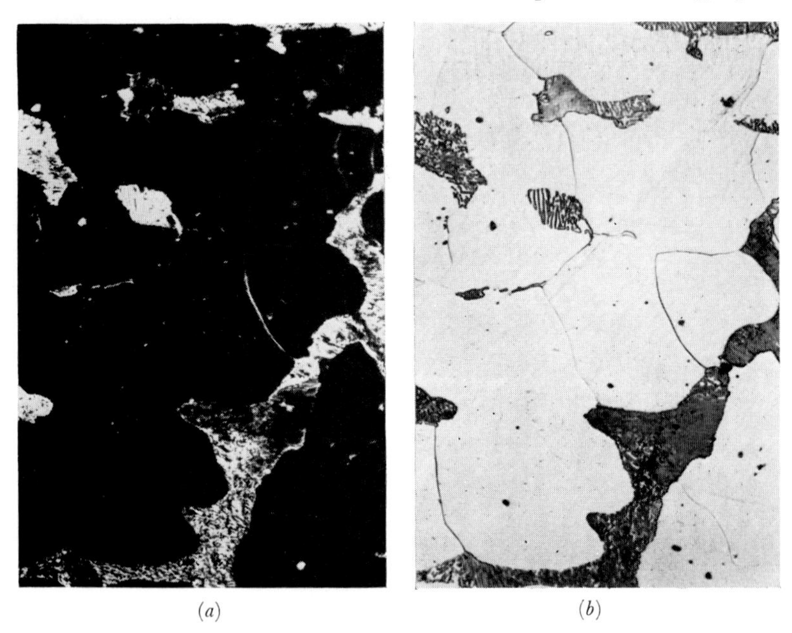

Fig. 30. (*a*) Dark-ground illumination ("oblique dark-ground") obtained using a catoptric condenser on a specimen of low carbon steel; etched, ×500.
(*b*) The same field, with bright-field illumination.

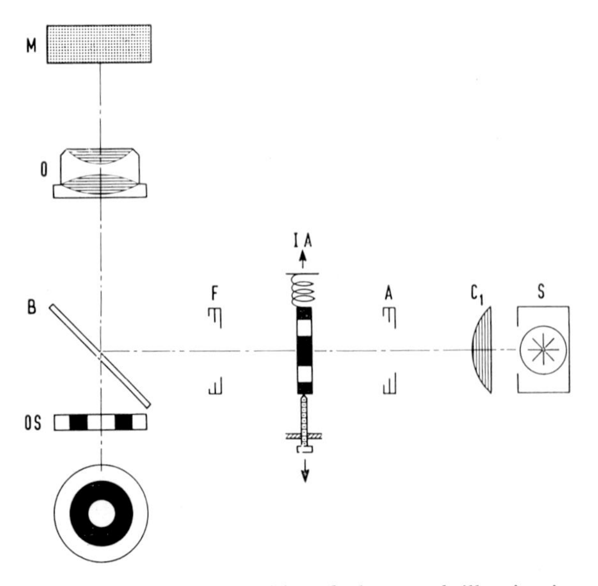

Fig. 31. Opaque-stop or sensitive dark-ground illumination; the arrangement to adapt to this form of illumination. An annular illuminating disc at IA is matched to the opaque stop OS (shown in plan below). The disc IA may be moved across the optic axis. See also Fig. 24 (*d*).

light from a plane specimen set normal to the optic axis; the design of the two stops is illustrated in Fig. 31.

If the illuminating stop IA can be moved across the beam it can be used to measure inclination on the surface of a specimen. A sloping area will deflect the direct beam so that it is no longer absorbed by the stop OS, and so change from true dark-ground to a Schlieren arrangement; movement of IA can then be made to return the image of the feature to dark-ground. The extent of this movement may be used, after suitable calibration, to measure the slope of the feature examined. The sensitivity of this arrangement is approximately thirty times that of normal illumination in detecting small changes of inclination; a sensitivity of 1° can be obtained at a magnification of × 200 using a numerical aperture of 0·45. This sensitivity is also, of course, obtainable using more conventional oblique or dark-ground techniques, but the opaque-stop arrangement has the advantage of being both symmetrical and adjustable. However, greater sensitivity and precision in measuring slopes can be attained with other techniques, to be described later, so that the opaque-stop microscope is probably best regarded as an inexpensive and relatively simple method of gaining qualitative information about changes of slope.

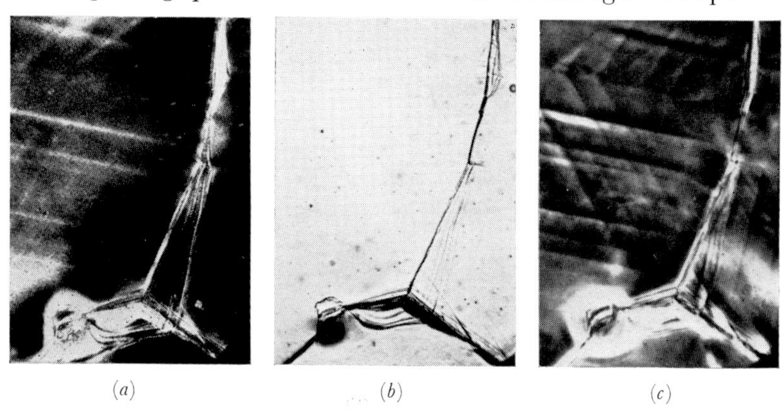

(*a*) (*b*) (*c*)

Fig. 32. (*a*) Opaque-stop, set for sensitive dark-ground, used on a creep specimen of lead, showing grain-boundary migration slip and polishing pits; unetched, ×200.
(*b*) Same field with bright field.
(*c*) Same field with positive phase-contrast.
(See also Fig. 65 of Chapter 6.)
(Reduced to $\frac{5}{6}$ in reproduction)

Fig. 32 (*a*) shows an example of its use to reveal the small inclinations on a specimen; Fig. 32 (*b*) shows the same area with bright-field illumination, and a further comparison is given in Fig. 32 (*c*) using positive phase-contrast.

STOP-CONTRAST

Another method of illumination using a stop has been described by Hallimond[3], who called his device "stop-contrast". This stop has the form shown in Fig. 33, comprising a central opaque stop surrounded by two opaque annuli, the second one extending to the outside area of the disc. The stop is placed at or near the rear focal plane of the objective lens, so that it affects both the incident and reflected light—see Fig. 24 (*e*). A stop of this kind can be made by photographing a drawing on paper at an appropriate reduction; or it can be drawn with indian ink

Fig. 33. Stop-contrast (Hallimond)—the form of the stop which is incorporated at the rear focal plane of a long-focus objective (> 16 mm). See also Fig. 24 (*e*).

on a cover slip rotated on a polishing wheel or gramophone turntable; slight inaccuracies of draughtsmanship are not important. The radii of the clear annuli are arranged so that light from the inner annulus reflected as the direct beam passes through the opposite side of this annulus when the specimen is flat and normal to the optic axis; some diffracted light then also passes through the outer annulus and a normal bright-field image is formed. As the surface of the specimen is tilted, more and more of the direct beam is cut off by the first opaque annulus so that the image becomes dark-ground. In this way changes of slope are revealed as gradations in intensity.

The necessity to place the stop at the rear focal plane of the objective lens restricts the use of this technique to lenses of long focus (16 mm or longer), because only these have the rear focus sufficiently outside their rear components. Because the incident beam has to be limited to the diameter of the inner clear annulus, the full aperture of the lens is not used; the size of the beam is limited, of course, by closing the aperture stop. However, long-focus lenses are not of high N.A., so that they would not be used at all if maximum resolution were being sought.

Fig. 34 (*a*) shows a micrograph taken using stop-contrast and Fig. 34 (*b*) shows the same area with bright-field illumination; in Fig. 34 (*c*) we see the effect of partially closing down the aperture stop; Fig. 34 (*d*) is an interferogram.

Phase-contrast Techniques

The group just described under the general heading of "interference-balance" methods have all been based on modifications to the amplitudes of the direct and diffracted beams and the relative proportions of the

Fig. 34. (*a*) Hallimond stop-contrast used on a creep specimen of a lead-thallium alloy; unetched, $\times 150$.

(*b*) The same area with bright-field illumination.

(*c*) The same area with bright-field illumination, aperture stop closed.

(*d*) The same area with multiple-beam interference fringes (see Chapter 6).

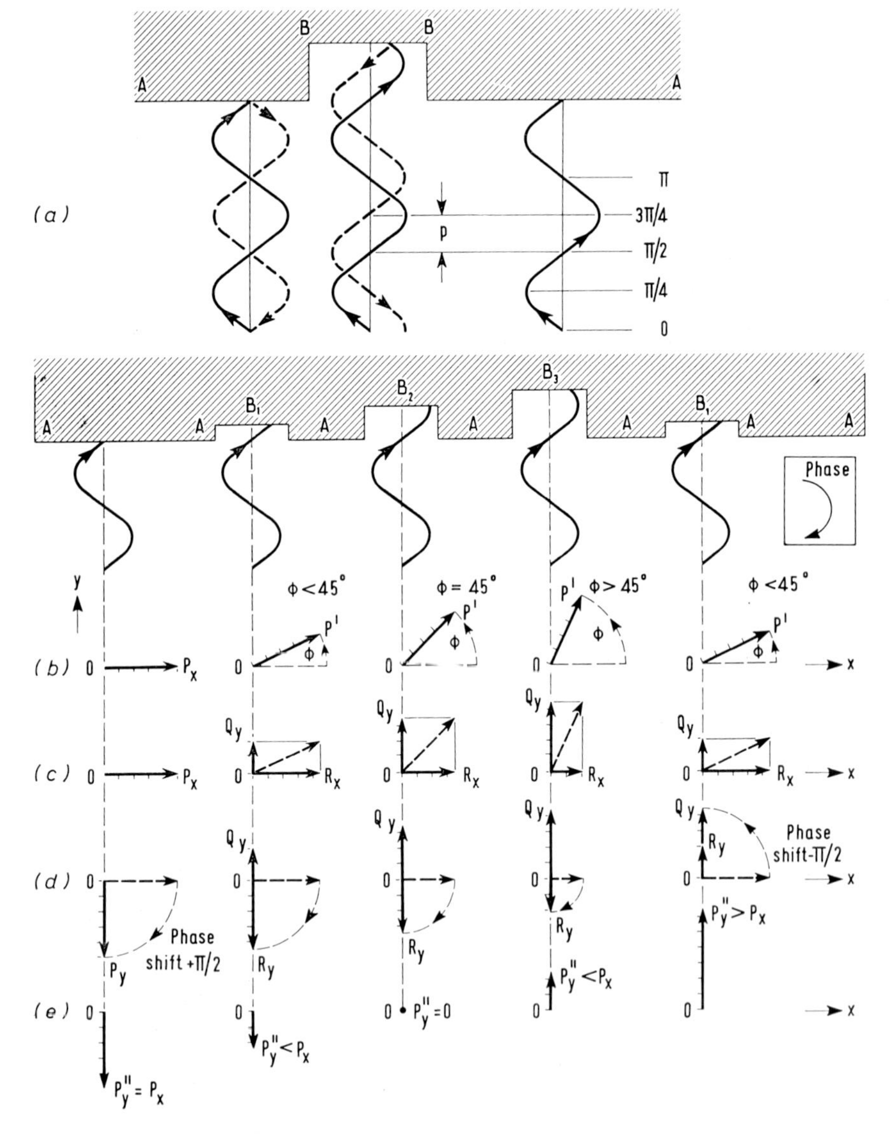

Fig. 35. (*a*) Showing how a phase difference *p* arises between beams reflected from different levels AA and BB of a uniformly bright specimen.

(*b*) to (*e*). Illustrating a simplified theory for the formation of phase-contrast images, from a series of levels A, B_1, B_2, and B_3. The first column represents positive phase-contrast for level B_1, the last column represents negative phase-contrast for this level.

The reflected light is represented by a vector OP in (*b*), and this is resolved into direct and diffracted beams in (*c*); in (*d*) the direct light is advanced or retarded by $\pi/2$ (through the agency of the phase ring) and the final intensity of the image is given by the recombined vectors in (*e*).

two kinds of beam. If we consider a uniformly reflecting specimen which has small localized changes of level (Fig. 35 (*a*)), bright-field illumination will not give much (if any) contrast in the image. However, the light does carry a message which is related to these changes in level, because in travelling, for example, to and from level B it will be retarded in phase compared to light reflected from level A. The eye cannot detect changes in phase of light and so a special technique has to be employed to convert phase differences to amplitude differences.

The necessary trick was devised in 1934 by Zernike[4], who used it to reveal phase differences in transmission microscopy, where the changes in phase are associated with changes in refractive index. The adaptation for reflected light was made in 1948 by Cuckow[5] and equipment for phase-contrast illumination is now available for all metallographs and many bench microscopes.

The principle on which the operation of phase-contrast is based can perhaps best be appreciated by considering a treatment derived from a simplified theory, due to Köhler and Loos[6]. This theory, in fact, contains steps which are false, but gives some explanation for the conversion of phase differences into amplitude differences, a phenomenon only explicable otherwise through the use of wave equations, or through a rather lengthy exposition such as that given by Françon[7] (p. 66 *et seq.*)

Suppose a uniformly reflecting specimen to have small areas with small changes in level, such that the light incident on them normally travels further to the depressions B_1, B_2, etc. than to the general level A (Fig. 35 (*b*)). Represent the reflected light from A by the vector $\mathbf{OP}_x$ and from B_1 by $\mathbf{OP}'$. By this we mean that the amplitude of the light is represented by the length of the arrows $\mathbf{OP}_x$ and $\mathbf{OP}'$, and thus the intensity by the square of the numbers characterizing these lengths. Also, the phase of the light is represented by the angle between the arrow and some reference direction, which has been taken here to be the horizontal. Because the reflectivity of the two levels is the same, the length $\mathbf{OP}_x = \mathbf{OP}'$ but because of the phase difference arising from the extra distance of travel to B_1, $\mathbf{OP}'$ lags behind $\mathbf{OP}_x$ by an angle ϕ. In Fig. 35 three cases are shown with B_1, B_2 and B_3 corresponding to $\phi < 45^\circ$, $\phi = 45^\circ$ and $90^\circ > \phi > 45^\circ$.

Since the depression B_1 is small it is assumed to be equivalent to an element of a diffraction grating and to diffract light so that in Fig. 35 (*c*) the vector $\mathbf{OP}'$ is replaced by $\mathbf{OR}_x$ along the direction OX, representing the direct light*, and $\mathbf{OQ}_y$ along a direction at right angles, representing the diffracted light. It is known from experiment that the diffracted light lags behind the direct light by $\pi/2$.

Now in Fig. 35 (*d*) suppose that by some means the direct light is advanced in phase by $\pi/2$ without affecting the diffracted light. Thus $\mathbf{OP}_x$ is moved to $\mathbf{OP}_y$ and $\mathbf{OR}_x$ to $\mathbf{OR}_y$. In Fig. 35 (*e*), to represent the resultant intensities, we combine $\mathbf{OQ}_y$ and $\mathbf{OR}_y$, so that the diffracting

* This appears to be an arbitrary step, but can perhaps be justified on the grounds that the process is one of referring the light from B_1 to that from A as a basis for comparison.

level B_1 is now represented by $\mathbf{OP}''_y$. It will be seen that $\mathbf{OP}''_y$ is less than $\mathbf{OP}_y$ when $\phi < 45°$, so that the depression appears less bright than the general level. However, making the same changes for B_2 and B_3 we find that when $\phi = 45°$ the contrast is greatest and as ϕ increases towards $90°$, the contrast decreases again. As ϕ continues to increase the contrast will vary cyclically every $\pi/2$.

If the same set of operations were performed but with the direct beam retarded by $\pi/2$, the intensity of light from B_1 would then be found to be greater than from A; maximum contrast is again found occurring at $\phi = 45°$. A phase difference of $45°$ is, of course, equivalent to $\pi/8$, so that for green light we would expect the maximum contrast for a depression of 675 Å, provided that the intensity of the direct component of the light is unaltered by its passage through the phase-changing annulus. In practice, as will be described later, there is an absorption of the direct light, specifically arranged to increase sensitivity, and so the maximum contrast is shifted to a value of 250–300 Å. It can readily be seen in Fig. 35 how absorption of the direct light would bring about this change, for it would be represented by shortening $\mathbf{OR}_x$ so that $\mathbf{OR}_y$ becomes equal to $\mathbf{OQ}_y$ at a smaller value of ϕ.

EQUIPMENT

In basic phase-contrast microscopes the means is provided to accomplish either or both of these operations of phase changing. The essential equipment consists of a disc to provide a hollow cone of rays, as in the opaque-stop microscope (IA in Fig. 36 (*a*)) and a *phase ring* which is placed at PP or some other conjugate point, often the rear focal plane of the objective.

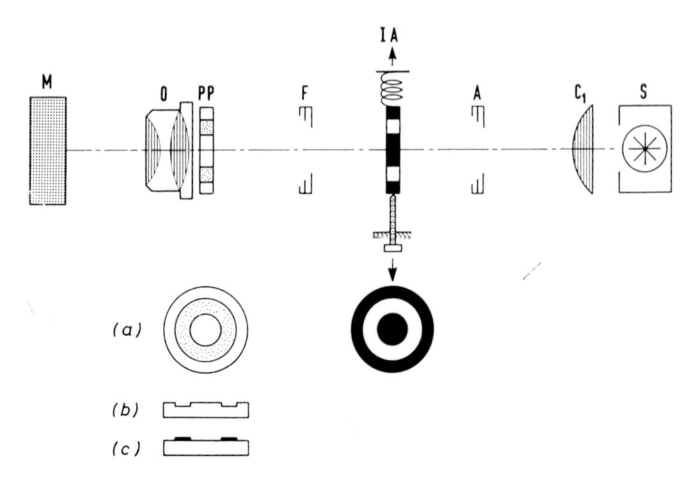

Fig. 36. (*a*) Arrangement for phase-contrast microscope. The illuminating annulus IA is capable of movement in two directions across the optic axis and is matched by a phase plate PP, often placed at the rear focal plane of the objective, but which may be at some conjugate point to this; e.g. just below the beam-splitter, as in Fig. 31. Two forms of phase plate are shown, the grooved type in (*b*) and one having a deposited annulus in (*c*).

The phase ring is a disc having an annulus of such a size that all the direct light from a flat specimen placed normal to the optic axis, and illuminated by the annulus, passes through it. This phase annulus is made so that it can either advance or retard the direct light by $\pi/2$ with respect to the diffracted light; since the diffracted light is bent away from the central beam most of it is transmitted through other parts of the phase-changing disc and therefore does not suffer the phase change. Fig. 24 (*f*) shows the path of one half of the rays in a phase-contrast arrangement, representing the imposed phase difference between direct and diffracted beams by the distance shown; the similarity to the opaque-stop microscope should be noted.

The phase disc can take one of two forms. It may be a glass disc in which a groove has been cut, Fig. 36 (*b*), so that the phase-changing is provided by the shortening of the distance the direct rays have to travel through the medium of higher refractive index (i.e. the glass). Alternatively, it may be an annulus of a suitable material deposited on the glass disc, Fig. 36 (*c*), so providing a longer path through a medium of higher refractive index than air. Magnesium fluoride is one material which is used to do this. Both kinds of phase ring also have a deposit of an absorbing material, usually antimony, on the annulus; this is provided to cut down the intensity of the direct light and so permit more complete interference with the diffracted beams. It has been suggested by Burch[8] that a good deal of the contrast effect commonly observed in the metallurgical phase-contrast microscope is probably attributable to this absorption, producing, in effect, amplitude contrast.

By controlling the depth of the groove, or the thickness of the deposit, the phase of the direct light may be advanced or retarded. When it is advanced by $\pi/2$ it is known as positive phase-contrast and this, following Fig. 35, shows elevations bright, depressions dark. Retarding by $\pi/2$ gives negative phase-contrast and reverses the contrast shown by the positive setting, again in accordance with Fig. 35. It should be realized that phase changes of other intermediate values would also give image contrast but, as Fig. 35 suggests, changes of $\pi/2$ give the maximum contrast. Indeed, there are instruments designed to give variable change of phase, and there are operational advantages in being able to do this (see "Variable Phase-contrast").

OPERATION

In order to use phase-contrast equipment, several adjustments have to be made to the microscope. The illuminating annulus is inserted and sometimes an auxiliary lens has to be put near the field condenser C_2 to focus the image of the annulus on to the plane of the phase disc (after reflection from the specimen). The phase disc is often built into the objective lens; its presence when using the objective with normal illumination can be ignored except in extremely critical applications. Alternatively, it may be carried on a slide which is placed just below the beam-splitter (Fig. 31). In some cases a single illuminating annulus may be used to illuminate phase rings of a size appropriate to each objective; in other cases the illuminating annuli have to be interchangeable to match the phase rings.

On some microscopes it is possible to change quickly from phase-contrast to bright-field illumination, for example by opening a stop at the illuminating annulus, so that light then passes through a second, outer, clear annulus and gives bright-field conditions. It is a distinct advantage to be able to change from the one kind of illumination to the other quickly, so that continual reference can be made to the more familiar bright-field image. Another modification consists of a triple annulus with a stop that clicks into position at settings which limit the light in the following way. When this stop is fully open, light passes through all three clear annuli, including the largest outer one; the illumination is then bright-field. On closing to the first position, the image is partly phase-contrast, from the inner annulus, and partly bright-field from the middle annulus which is of comparable area to the inner one. On closing to the second position, only the inner annulus operates; this limits direct light to a phase-changing ring and true phase-contrast results.

In general it is not sufficient to focus on to the specimen and insert the phase ring and illuminating annulus to obtain phase contrast. This is because the surface of the specimen over the area examined will not always be accurately normal to the optic axis, and consequently the direct light will not all pass through the phase ring on reflection. To

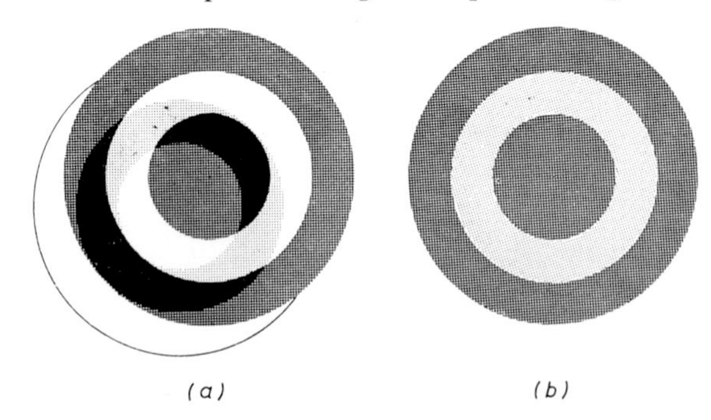

Fig. 37. Adjustment of the phase ring and illuminating annulus to give phase-contrast illumination. In (*a*) the direct light does not all pass through the phase-changing annulus (dotted); the correct alignment is shown in (*b*).

adjust this, the specimen is focused, the ocular is removed and replaced with an auxiliary lens (often called a Bertrand lens) which is focused on to the phase disc. An image of the clear annulus of the illuminating disc will then be seen as in Fig. 37 (*a*) and the disc is moved by adjusting screws, or the level of the specimen altered, until the image of the annulus coincides with that of the phase ring—Fig. 37 (*b*). The sizes of the two are usually made so that the phase ring is slightly larger than the image of the illuminating annulus. With an undulating specimen it may be necessary to repeat this registration frequently as the specimen is scanned.

It also often happens that the specimen is sufficiently rough so that there are several images of the illuminating annulus, one from each of the several facets being viewed. In this case, of course, true phase-contrast can at the best only be obtained for each feature in turn but may not be attainable for any feature at all.

VARIABLE PHASE-CONTRAST

A device has been developed by Françon and Nomarski[9] which can continuously vary the phase difference between the direct and diffracted light from $\pi/2$ advancement to $\pi/2$ retardation. The equipment (Fig. 38)

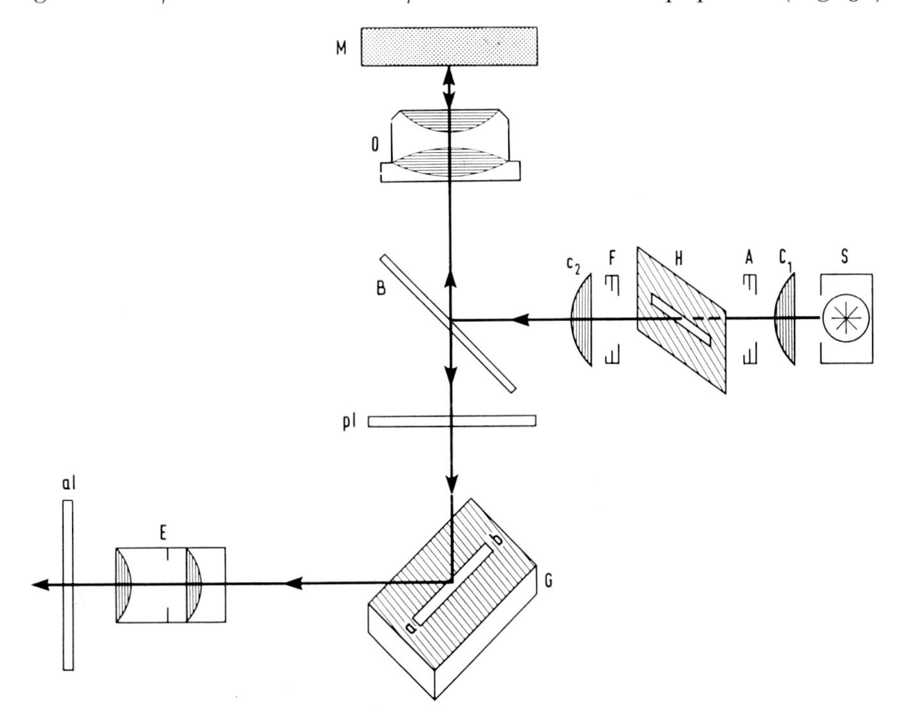

Fig. 38. Arrangement for variable phase-contrast (after Françon and Nomarski). The usual illuminating annulus becomes a slit H and is matched to a strip on a glass block G; the phase change depends on polarization on reflection at this block.

uses a slit illuminator H, instead of an annulus; an image of this slit matches a clear strip (ab) on an aluminized glass block G which is placed at a conjugate plane to the eyepiece, but in such a way that the direct light reflected from (ab) is completely polarized (see Brewster's Law in the chapter on polarized light). The diffracted light is totally reflected from the metallized layer.

The thickness of the metal is chosen, in conjunction with the angle of incidence, to give the $\pi/2$ phase difference associated with normal phase-contrast procedures. A polarizer (pl) is set, as shown, below the beam-splitter with its plane of polarization at 45° to the plane of incidence of the phase plate G. Rotation of the analyser (al) controls the degree of phase change at G from $+\pi/2$ to $-\pi/2$. With monochromatic light the resulting images are exactly those expected from normal phase-contrast, but with white light there is selective elimination of successive wavelengths from the white light, and the image is then rendered in colour contrast.

The device is somewhat limited by the reduced N.A. and directional nature of illumination which are both results of the use of the slit illuminator.

SENSITIVITY AND USES

With a phase change of $\pi/2$ the sensitivity of the phase-contrast microscope is such that a change of level of 50 Å can be detected. As pointed out previously, maximum contrast is found for a change in level in $\sim$ 250 Å with a phase ring of typical absorption. These figures pertain to light of 0·5 μm wavelength. Above $\sim$ 500 Å (with green light) the contrast can only be ambiguously interpreted, because the phase difference introduced by the object then exceeds the critical value of ϕ. Steps of size greater than this may then be characterized by fringe-like reversals of contrasts. Generally, however, the features of interest are sufficiently complex for the fringes to become smeared out, thus possibly assisting in the enhancement of contrast. Alternatively, the features are sufficiently simple for the fringes to be readily recognized as such. Sometimes fringes which are clearly seen, for example, on the faces of a hardness impression, turn out to be an effect of using an annular illuminating stop alone, and merely to be made clearer when the phase plate is used. Fig. 39 shows this effect;

(*a*)

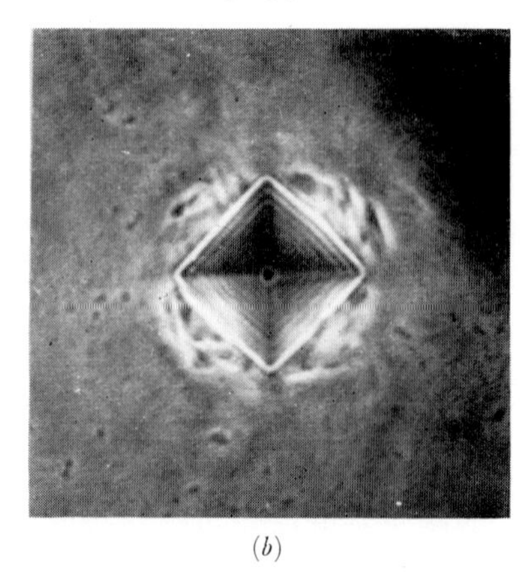

(*b*)

Fig. 39. Showing that fringes on the faces of a pyramidal hardness impression are (*a*) present when the annular illuminating stop alone is used, and (*b*) enhanced in contrast with the phase plate in position.

clearly, in this case, the contrast seen does not mean that the faces of the hardness impression are corrugated.

Changing from positive to negative phase-contrast should reverse the contrast of an image where the intensity arises from true phase-contrast;

this points to an obvious advantage in having both varieties of equipment readily available. An example is shown in Figs. 40 (*a*) and (*b*).

As might be expected, phase-contrast is much more likely to give new information about a deformed specimen where the structure consists of steps, folds, etc. than about a typically etched specimen, where the structure is characterized by texture or coloration. The detection of a very fine slip formed during creep (Fig. 41) provides one of the best

(*a*) (*b*)

Fig. 40. Reversal of contrast at suitable features on changing from (*a*) positive to (*b*) negative phase-contrast.

The specimen is of titanium (contaminated with oxygen) and quenched in the arc furnace, so producing a Widmanstätten structure; unetched. × 500.

Fig. 41. Fine slip formed close to a sliding grain boundary in a lead bicrystal, revealed using positive phase-contrast. The broad slanting band is the grain boundary, where a step too large for phase-contrast has been formed, and on which marks due to migration are present.

examples of the use of phase-contrast to detect a feature which would otherwise have been found only by the use of the electron microscope. In many other examples which have been published it is likely that phase-contrast gives good, possibly the best, overall realization of the features of interest but is not unique in its ability to detect them optically. Quite often, oblique illumination, sensitive dark-ground, or even the "out-of-focus" technique might be capable of revealing aspects of the structure according to the particular setting of the illumination used. An example has already been given in Fig. 29, showing how only some slip systems are revealed, although quite clearly, with one particular direction of oblique illumination. It is often, therefore, much easier to scan a specimen using phase-contrast illumination instead of one of the other techniques where continual adjustments have to be made to the conditions of illumination or focus. However, it should be noted that such adjustments may also be required to maintain a phase-contrast image when the specimen surface is so rumpled or faceted that the direct beam is deflected to pass outside the phase ring. In this case the advantage of using phase-contrast is much diminished or even lost.

The ability of phase-contrast illumination to reveal very small features makes it a sensitive way of showing up polishing scratches or the remains of scratches; this, of course, is not always an advantage to the metallographer. Fig. 42 (*b*) shows polishing scratches in the ferrite grains of a

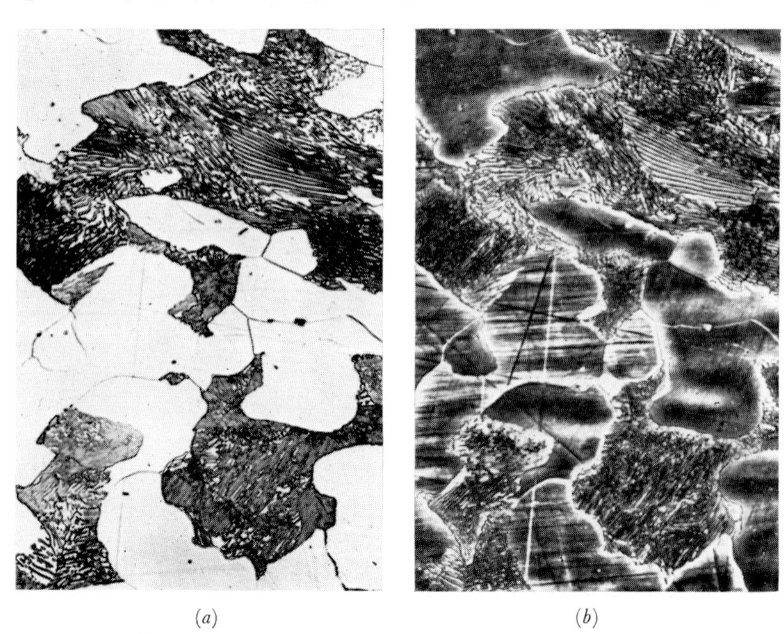

(*a*) (*b*)

Fig. 42. Pearlite and ferrite; etched. $\times 500$.
(*a*) Bright-field;
(*b*) Positive phase-contrast.
Phase-contrast does not add very greatly to the metallographic information except in that it reveals the inadequacy of the preparation of the specimen.

pearlitic structure and also suggests that the phase-contrast image of the pearlite itself is probably more confusing, rather than more informative, than the bright-field counterpart in Fig. 42 (*a*).

Perhaps the most general advantage of phase-contrast illumination, outside its use in studying deformation, is that it will reveal very lightly etched structures. The art of etching is a very empirical one (see Chapter 9), and it often happens that new alloys, particularly those based on metals like niobium, titanium, etc., are difficult to etch without using hot, strong acids. Over-etching, staining and pitting often result very readily. A very light etch, or the small amount of relief consequent upon even a good mechanical polish, will often suffice to give good realization of the

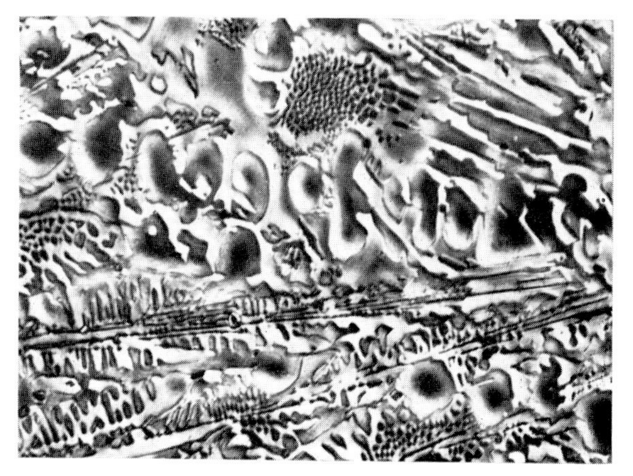

Fig. 43. Phase-contrast revealing the structure of an unetched specimen of "Ni-hard" cast iron. × 100.

structure when phase-contrast is used. An example has already been given in Fig. 40, and Fig. 43 shows another.

A use for phase-contrast equipment which occasionally can be of assistance is that of employing the registration of the image of the annulus and the phase ring (Fig. 37) to level a specimen accurately. This remark does not, of course, have any point in connection with inverted-stage microscopes.

Conclusions

Each of the techniques described has its special uses, and ideally the metallographer should be able to produce the best method for each problem as it arises. This means that none of this group of techniques is a panacea for all the problems associated either with the attainment of adequate contrast or with the revelation (as distinct from the resolution) of detail. Moreover, in practice they are essentially qualitative in their ability to assess changes in surface level; if quantitative information is required, interferometry, taper sectioning or some related technique should be used (see Chapter 6).

The extra cost of the various pieces of equipment required for the group of techniques described in this chapter is small compared with that of the microscope itself. The most expensive is that for phase-contrast, especially if both positive and negative types are acquired. At the other end of the range, the out-of-focus technique costs nothing. Oblique illumination is also simply attained, and there is considerable merit in having the aperture stop mounted on a carrier which can be moved across the illuminating beam and rotated to give various degrees and directions of oblique illumination. When phase-contrast equipment is available, it is well worth making the appropriate opaque stop to give opaque stop-contrast, thus inexpensively adding a flexible technique which combines oblique and dark-field types of illumination.

Hallimond's stop-contrast is so easy to make that its description seemed warranted in this chapter, even though the ready availability of phase-contrast now makes it outmoded; this was not the case when Hallimond proposed the method about twenty years ago.

Perhaps all of this group of techniques other than the out-of-focus one are now being superseded by the interference-contrast technique of Nomarski (described in Chapter 6). This is able to give a series of effects which merge into one another and ranges from the equivalent of phase-contrast to that of dark-field illumination but also includes varied (and beautiful) colour effects. However, Nomarski interference-contrast involves the use of polarized light and this may sometimes be a disadvantage, because polarizing effects complicate the information given by the image and reduce its absolute intensity compared to conventional phase-contrast illumination. Even conventional phase-contrast may demand rather long exposure times on the comparatively slow plates used in metallography.

REFERENCES

1 M. Berek, *Optik*, 1945, **5**, 144 and 329.
2 W. M. Lomer and P. L. Pratt, *J. Inst. Metals*, 1951–52, **80**, 409.
3 A. F. Hallimond, *Nature*, 1947, **159**, 851.
4 F. Zernike, *Zeit. f. Tech. Phys.*, 1935, **16**, 454.
5 F. W. Cuckow, *J. Iron & Steel Inst.*, 1949, **161**, 1.
6 A. Köhler and W. Loos, *Naturwiss.*, 1941, **29**, 49.
7 M. Françon, *Progress in Microscopy*, Oxford, 1961 (Pergamon).
8 C. R. Burch in *The Examination of Metals by Optical Methods*, London, 1949 (B.I.S.R.A.).
9 M. Françon and G. Nomarski, *Rev. Opt. Theoret, et Inst.*, 1950, **12**, 619.

6 Interference Techniques

The variations in image described in the preceding chapter depend upon manipulation of the interference of rays diffracted from the specimen; in this chapter we shall consider a group of techniques in which the interference is between one set of rays reflected from the specimen and another reflected from some reference surface. Sometimes the reference surface is ingeniously arranged to be an adjacent part of the specimen itself. The result of such arrangements is to produce patterns of fringes which contour the specimen.

Two-beam Interference

The simplest arrangement is typified by the conditions for production of *Newton's Rings* (Fig. 44) A spherical lens L is placed on a flat glass plate

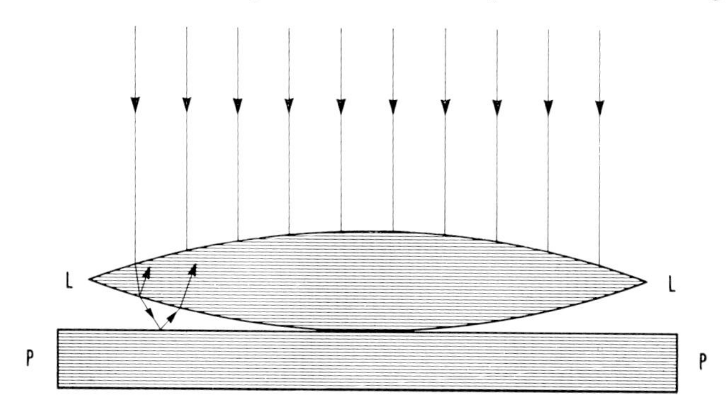

Fig. 44. Production of Newton's Rings by interference between beams reflected at the lower surface of lens L and the upper surface of a plate P. These beams reinforce or cancel according to the path difference between them, and so produce concentric dark and light fringes.

P and illuminated by a beam of monochromatic light at something like normal incidence. The beam is partly reflected from the lower glass-air interface of the lens and partly from the air-glass interface of the plate. The amplitudes of these two reflected beams are approximately equal but phase differences are introduced according to the difference in their path lengths; a series of concentric fringes is therefore produced by interference between various pairs of beams, and these represent contours of the lens referred to the level of P. The spacing of these fringes depends both upon

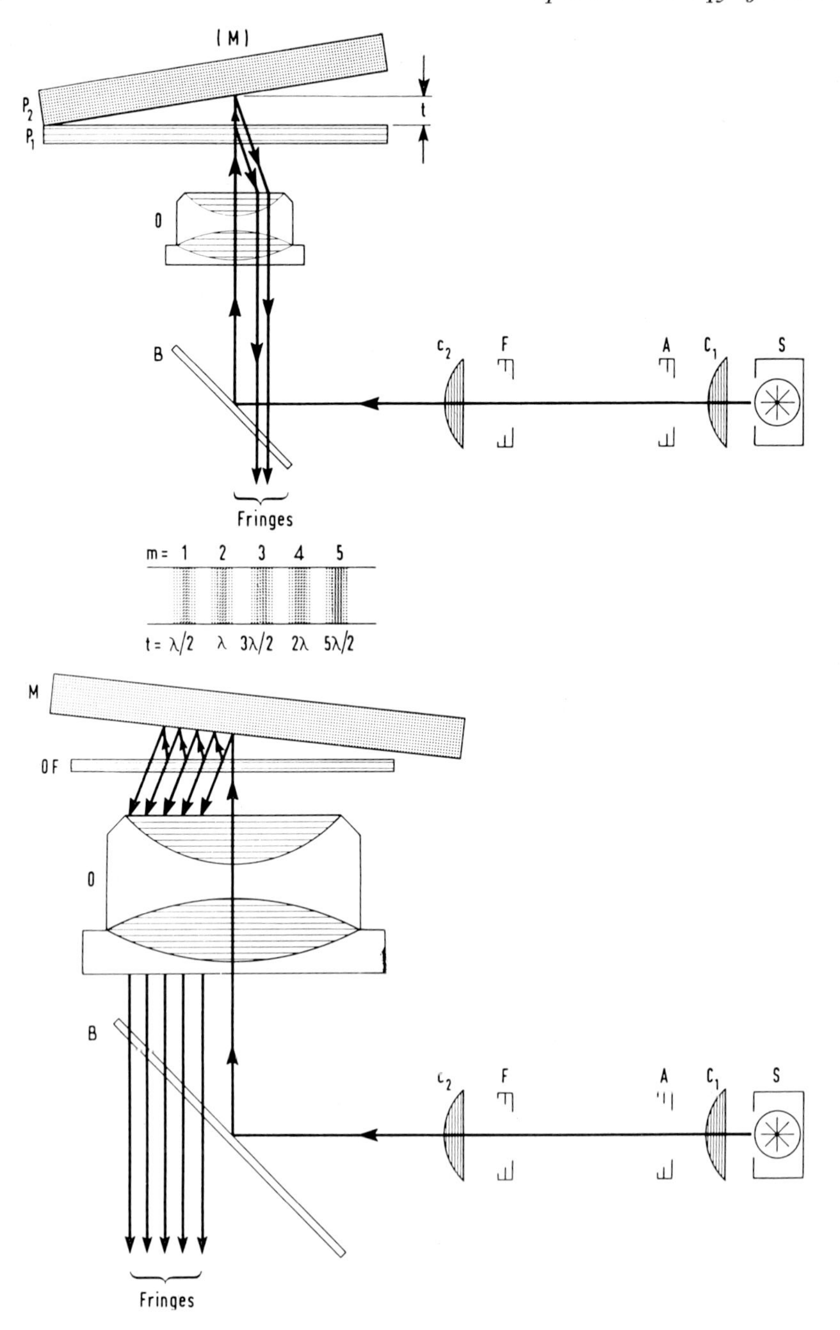

Fig. 45. (*a*) Two-beam interference on a metallurgical microscope. The specimen M replaces the lens L in Fig. 44 and a thin flat plate P_1 becomes the reference plate P. The separation t between M and P_1 gives rise to fringes according to the number of half-wavelengths in the path difference.

(*b*) A similar arrangement with conditions such that multiple reflections occur, producing multiple-beam interference fringes.

the wavelength of the light and the angle of the "wedge" (of air) between the two pieces of glass.

Straight fringes would, of course, be observed if the lens were replaced by a second piece of plane glass inclined at a small angle to the lower glass. Suppose two such plates to be set up in this way (Fig. 45 (*a*)) and illuminated with a collimated beam of monochromatic light through the usual illuminating train. The fringes are viewed after transmission through the beam-splitter B. Let the refractive index of the medium between the plates P_1 and P_2 be n, the wavelength of the light employed λ, and the incident beam be normal to P_1, so that it is also very nearly normal to P_2, and let the wedge thickness at a particular point be t. It should be noted that the angle drawn in Fig. 45 (*a*) is much larger than it would be in practice. Then the position of a fringe of order m is given by

$$m = 2nt/\lambda \tag{1}$$

The word "order" simply refers to the numbering of fringes starting from the one at the smallest value of t which gives rise to a fringe. The fringes appear to be localized in the "wedge" of air, and because $n = 1$ for air, each fringe represents the locus of points for which t is constant;

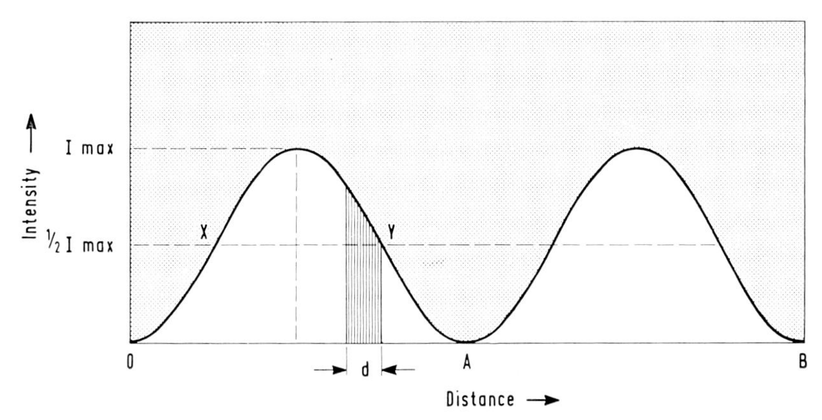

Fig. 46. Intensity in a two-beam system of fringes, such as would be produced by the arrangement in Figs. 44 or 45 (*a*). The fringe spacing AB is equivalent to a change in t of $\lambda/2$ and can be measured with an accuracy defined by $d \simeq 1/5$ of $\lambda/2$.

these light and dark areas are of equal width. It can be shown that the intensity of the light in the fringes follows a $(\cos)^2$ curve, Fig. 46. On moving from one fringe to the next, the value of t—the thickness of the wedge or air space—has been changed by $\lambda/2$.

Half the width of the fringe is defined as XY, which is the width of the intensity curve at half the maximum value of intensity. The distances between orders is clearly AB = 2XY. Suppose that the position of a fringe can be located to one-fifth of the half width (that is, within the shaded area in Fig. 46), then the precision with which t can be measured is $\lambda/20$. Tolansky quotes a value of $\lambda/40$ as being often accepted in the literature, but suggests that it is rather optimistic. For worked glass surfaces the precision should be between the two values we have quoted. For metals, which we will consider presently, even $\lambda/20$ is rarely attainable, because of the relative roughness of a metal surface compared to that of glass.

If we now replace P_2 by a plate with an undulating or roughened surface but retain P_1 optically flat, the fringes will again follow regions of equal wedge thickness, or of the same value of t, and thus will become contour lines plotting the surface levels of P_2 with a height between contours of $\lambda/2$. By gently pressing P_1 and P_2 together it will be found that the fringes move away from the smallest value of t. Thus for a roughened specimen, pressing the plates together causes the fringes to move away from the places where the separation between the plates is least.

The arrangement in Fig. 45 can be adapted for use in a metallograph with P_2 now the metal specimen and P_1 a suitable reference plate. This plate has to be such that it can be interpolated between the objective and the specimen; alternatively, the focal length of the objective must be sufficiently long to accommodate the plate P_2. This kind of arrangement has been used successfully (Chalmers[1]), but has now been superseded by more sophisticated two-beam interference microscopes, or by multiple-beam techniques.

Two-beam Interferometers

There are several two-beam interference microscopes or attachments available. The principle of one of these is illustrated in Fig. 47. This instrument is known as a Linnik interference microscope. The specimen P_2 and the reference plate are widely separated, and the intensity of the beam from the reference plate P_1 matched more closely to that from P_1 than is the case when a plane glass plate is used as reference. This is important, for the visibility of the fringes is very dependent upon full interference between the two beams; any excess light in either beam will merely degrade the fringe contrast. By separating the reference plate from the specimen, it becomes possible to use a metal (or metallized) surface for it and so achieve the required matching. The price that has to be paid to do this is high, for it will be seen from Fig. 47 that a second objective O_1 has to be incorporated in the system to view the reference plate P_1, the normal objective O_2 viewing the specimen P_2. Illumination of both parts of the system is from the same source S, as it must be for sufficient coherency to exist to produce interference; the beam-splitter B reflects one beam to the specimen and transmits another to the reference plate. After reflection at P_1 and P_2 the beams are combined again after reflection and transmission at B, so that interference may occur. The optical paths from the beam-splitter to P_1 and P_2 and back again have to be matched, so that they are identical within a tolerance set by the wavelength of light. Hence the two objectives O_1 and O_2 have to be a good deal more than nominally identical. For example, rays which pass down one side of the lenses on their passage to either specimen or reference plate return along the other side after reflection; thus any aberrations in the lenses must be neglible or else similarly asymmetric in the two objectives, so that each beam is distorted identically if at all.

The position of the reference surface P_1 must be located accurately in order to attain matching of the optical paths BP_1, BP_2. This can be done fairly precisely by mechanical means—on one microscope the reference

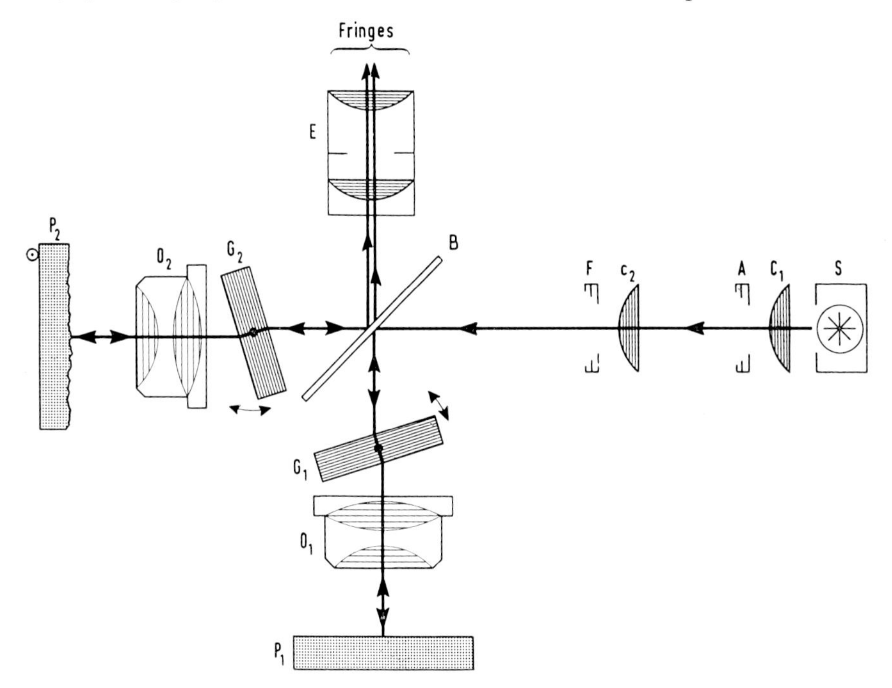

Fig. 47. A two-beam interferometer for use with reflecting specimens, known as a Linnik-type interferometer. The specimen P_2 and the reference plate P_1 are viewed by separate matched objectives O_2 and O_1. Fringe spacing and inclination are controlled by movements of the glass blocks G_1 and G_2.

plate is mounted in a cap which pushes on to a tapered holder, being positively located by the shoulder of the holder. Final adjustment is provided by the inclusion of a rotating glass plate G_2 in the path of the rays to the specimen; as the plate is turned, the optical path is changed by traversing more or less of the medium of higher refractive index, and so made to match the path to the reference mirror. The other glass block G_1 serves a different purpose and may be tilted about an axis normal to the plane of the diagram or rotated about the optic axis; these movements serve to give the equivalent of a tilt or rotation of the reference plate P_1, thus changing the "wedge" angle and its orientation. This makes it possible to close up or widen out the fringes, or to change their direction with respect to particular features. Such an adjustment is of great assistance and any equipment or alternative technique (like multiple-beam interferometry) which does not have this facility, or which achieves it by less direct and flexible means, is at a serious disadvantage in metallographic work.

The fringe contrast is best when the intensity of the two beams is about equal. Polished metals vary considerably in their reflectances and it is therefore desirable to be able to control the intensity of light reflected from the reference plate. The simplest method is to have two or three mirrors of varied reflectances which can be readily interchanged. With the Linnik instrument a complete set of mirrors of varying reflectance is required for each objective, because of the differing focal lengths of the objectives.

The primary magnification on the Linnik instrument can be made relatively high. One commercial instrument (Fig. 48) has three pairs of objectives giving × 10, × 25 and × 60, so that a final magnification of

Fig. 48. The Zeiss-Linnik interferometer, for which Fig. 47 is the schematic arrangement. The notation used is the same as in Fig. 47.

up to × 480 may be obtained using a × 8 ocular. The same magnification can be obtained on 35 mm film carried in a normal camera body which clips on to the microscope.

Single-objective Interferometers

There are several interference microscopes, or attachments to microscopes, which use a modified objective to produce fringes by the interference of two beams. All depend upon the introduction of some optical equipment between the objective and the specimen, and so may demand long focus objectives and low apertures. Three systems known to be incorporated in commercial equipment will be described here, although only one of these has the facility—almost essential in metallography—for changing fringe spacing and orientation.

Dyson's interference objective consists of three glass blocks mounted between the objective and the specimen. The reference beam is obtained by reflection at a small central mirror P_1 (Fig. 49 (*a*)) and the optical

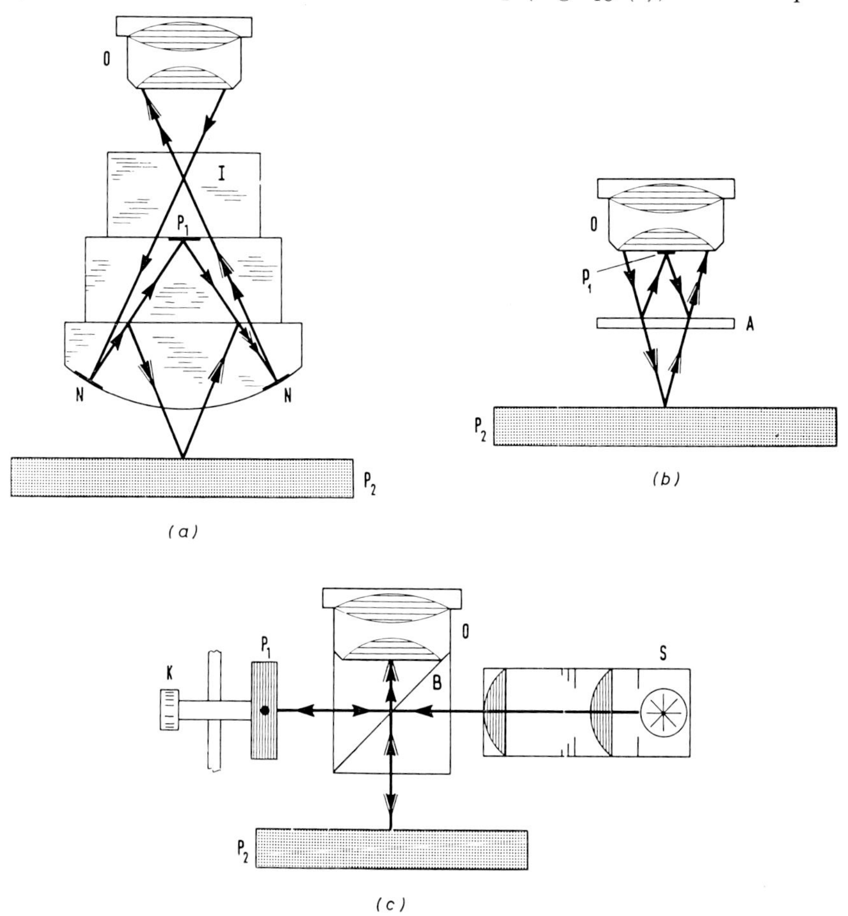

Fig. 49. Three types of two-beam interferometer which require one objective lens only (schematic):
(*a*) Dyson; (*b*) Mirau; (*c*) Watson.

paths of the direct and reference rays are matched by an internal reflection at a second, annular mirror N on the lowest of the blocks. By this means the fringes are formed within the upper block at I, and it is this fringe pattern which is viewed by the objective.

Mirau's interference principle is to place a semi-reflecting plate A between the objective and specimen and the reference mirror P_1 on the centre of the front component of the objective (Fig. 49 (*b*)). The optical paths to the mirror and specimen are balanced by the position of A and two-beam fringes are viewed by the objective.

Watson's interference objectives are designed to include a beam-splitter B between their front components and the specimen P_2 (Fig. 49 (*c*)), so that illumination has to be provided at this point by a special auxiliary source and collimator S. One beam proceeds to the reference mirror P_1 and the other to the specimen; after reflection these interfere to produce the fringes which are viewed by the objective. The mirror P_1 can be adjusted by movement of screws K to compensate for changes in level of the specimen and also to vary the spacing and orientation of the fringes. The primary magnification attainable with this arrangement is limited to $\times$ 20 with an N.A. of 0·6.

These three interference objectives or attachments give fringes which are of the sensitivity and contrast obtained with the more elaborate Linnik instrument, but do not give quite as high a lateral magnification. Also, as mentioned above, only the Watson system seems capable of modifying the spacing and orientation of the fringes directly. However, a tilting and rotating stage for the specimen could overcome this deficiency, provided that the movements were capable of being made small enough and without sloppiness and backlash. The Watson attachment, and probably each of the others, is designed to fit on to a normal bench microscope, so that a camera which is attached to the eyepiece tube is required for photographic recording of the fringes. Since fringe patterns may require rather long exposures, particularly when using monochromatic light, the stability of a bench microscope may sometimes be a problem. The single objective interferometers have a great advantage over the Linnik type in their lower cost, but this must be set against their individual degrees of inflexibility.

Interpretation of Two-beam Interferograms

The fringe patterns obtained using two-beam interference microscopes are simply interpreted, for they are contour maps of the specimen surface and give an immediate impression of the topography to anyone used to map reading. In the interferograms the height between contours is half the wavelength of the light used: usually green light is employed, so this height is of the order of 2700 Å (0·27 μm). Features of smaller size may be detected and measured, for they distort the fringe or draw it out into a streak. Fig. 50 shows a schematic representation of the development of contour lines on a specimen having a ridge and trough of overall height greater than two units of height and also a subsidiary trough of less than one unit. A series of planes is imagined parallel to "sea-level" and these are equally spaced at vertical intervals and intersect the specimen surface. The line of intersection generates the contour line. Of course, if the surface of the specimen is more steeply inclined with respect to the reference level (Fig. 50 (*b*)) the contour lines are more closely spaced.

However, when the slope is across the ridges and troughs in this way, they are not very clearly shown up. If the slope of the specimen is parallel to the ridge and trough feature, rather than across it, then this gives changes in the direction of the contours (Fig. 50 (*c*)) rather than variations in their spacing, and such features are clearly revealed. The smaller

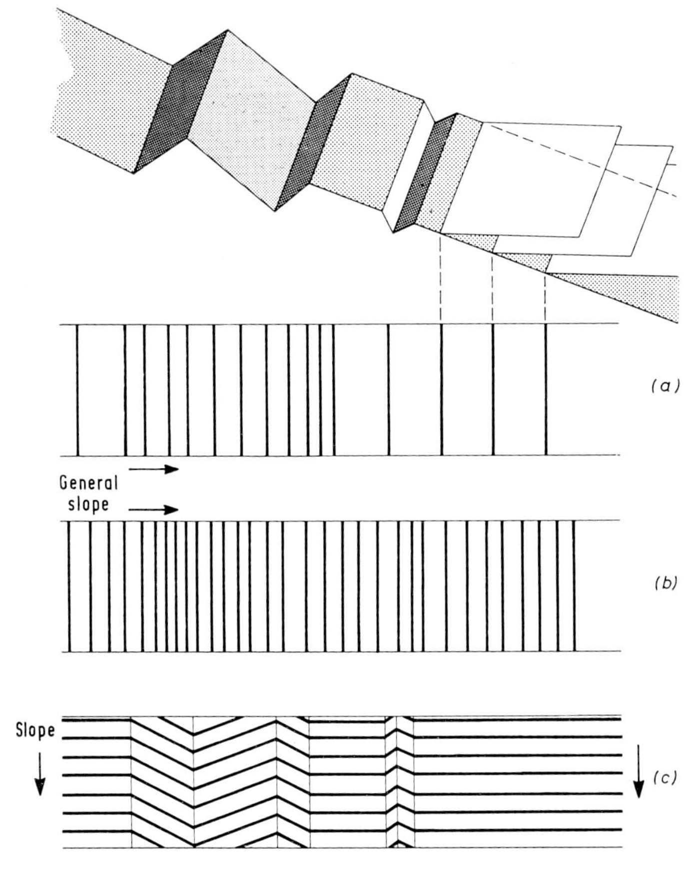

Fig. 50. Illustrating the fringe patterns produced by interference from simple ridge and trough features.
(*a*) Reference plate sloping gently across the ridges and troughs.
(*b*) Slope as in (*a*) but increased.
(*c*) Slope parallel to the features. Note that the diagram shows only the centre lines of the dark fringes, and is therefore more nearly equivalent to multiple-beam than two-beam fringes.

trough does not appear at all in Figs. 50 (*a*) and (*b*) but is revealed by the arrangement in Fig. 50 (*c*). In the more flexible types of two-beam interferometers, the "sea-level" is changed, by movement of the reference mirror, to effect the changes typified by Figs. 50 (*a*), (*b*) and(*c*). It can readily be appreciated that with complex topography it is very desirable that the means of making these changes should be speedy and positive.

So far we have spoken of the interpretation of the fringe patterns in a general and qualitative manner; this is very useful, but does not exploit the technique to the full. Suppose we have the fringe pattern shown in Fig. 51, which tells us that there is a step on the surface; then the height

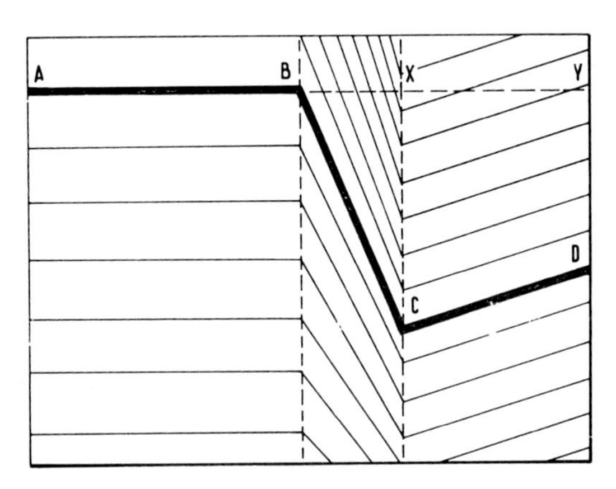

Fig. 51. Illustrating the manner in which the height of a simple step is measured using interference fringes. The number of fringes n between X and C is counted to measure the height at B; the height is then $n \times$ (half the wavelength of light used).

of the step can be estimated under normal conditions to one-tenth of a fringe spacing, i.e. to $1/10 \times 0{\cdot}27\ \mu m = 270$ Å or approximately 100 atom diameters. To do this, a particular fringe such as ABCD is taken as the reference and AB produced across the step to XY. The fringes on the right-hand side between BX and CD (excluding CD) are then counted, with an estimate of the fractional spacing between XY and the fringe below it; in the case shown (Fig. 51) this is 6·5 fringes, so the height of the step is $6{\cdot}5 \times \lambda/2$ where the wavelength of the light used is λ. It should be remembered that it is the number of fringe spacings that is required, so that counting should strictly be from fringe centre to fringe centre. An alternative procedure would be to count the fringes at B down the face of the step to X; often, however, this is difficult or even impossible, because the fringes are too closely spaced. For the most accurate work it is clearly an advantage to make these measurements on photographic prints, but where large numbers of readings are involved counts can be made on the microscope, the line XY being mentally constructed. The step could equally well have been measured by producing one of the right-hand fringes across to the other side and counting on the left of the step; indeed, this is a useful check, but it should be remembered that the height is measured at the point where the fringe is produced (geometrically). If the height of the step is not uniform (as in Fig. 51) the result of counting from AB produced across to XY will not necessarily be the same as counting from DC produced. Similarly, irregularities in the surfaces either side of the step could affect the measurement of the step height.

Fig. 52. Construction of a profile from an interferogram. (This is a multiple-beam interferogram of polished and lightly deformed aluminium; polishing pits and fine slip are present.) The line LR is drawn across the region of interest, and a perpendicular dropped from its intersection with each fringe. These lines change in length by a unit p from one fringe to the next; hence, the profile is built up. The magnification in depth is $m = 2p/\lambda$. Here $p = 1{\cdot}25 \times 10^{-1}$ cm, $\lambda = 5{\cdot}4 \times 10^{-5}$ cm, therefore $m \simeq 5 \times 10^{3}$. The lateral magnification is $\times$ 200.

When the feature has a height less than $\lambda/2$, this can be estimated by the fraction of the fringe spacing through which the fringe crossing the feature is distorted. This could be a gradual curve, or a sharp blip such as that shown in Fig. 50 (*c*). The former could arise for example, from a shallow pit (or hump) and the latter from a scratch.

PROFILES

Sometimes it may be of assistance in visualizing a feature to construct a profile. Having decided the section which is of interest, a line is drawn to represent the intersection of the imaginary section with the surface, for example LR in Fig. 52. At each intersection of LR with a fringe a perpendicular line is drawn, as shown in the figure, these lines being made one unit longer in turn on passing from one fringe to the next. The ends of the lines may then be joined up to give the profile. Whether the profile is moving up- or down-hill must be determined by some means (as described in the next paragraph). The magnification of the profile will, of course, be determined laterally by the objective and ocular used (plus any enlargement of the photograph) and the vertical magnification by the ratio of the unit lengths employed in the construction lines to $\lambda/2$; the magnification would be almost 10^4 if 2·5 mm (0·1 in.) represented $\lambda/2$ for green light.

Unless prior information is available as a guide, it is not easy to distinguish elevations from depressions or ridges from troughs. For small changes in height, the phase-contrast microscope should provide the answer; for larger changes shadowing effects with variations in oblique illumination may assist. Perhaps the most positive way of identifying the vertical relationship of adjacent features is to focus down from one level to the other with the fine focus, using an objective with an appropriately small depth of focus. With multiple-beam fringes another method is available.

Use of White Light

So far, the discussion has been of fringes produced using a monochromatic source, and these should be used wherever the highest precision is required; they also have an advantage in that they are visible over the whole of the field viewed. A system of fringes corresponding to part of that produced with monochromatic light results from the employment of an unfiltered white light as source. These fringes are coloured, each dark band of the monochromatic series being replaced by a specific spectrum of colours. The fringe spacing is then 3×10^{-5} cm, but greater sensitivity is obtained using monochromatic (thallium) light, when the spacing is $2{\cdot}7 \times 10^{-5}$ cm. This results in there being a clearly distinguishable central fringe, which is dark purple-black in its centre, shading off to blue on one side, brown on the other. The dark fringe can be used to identify fringes either side of a change in level across which it is too steep to follow the individual fringes, as in the example shown in Figs. 53 (*a*) and (*b*); also shown in this series is the same field with the fringes more closely spaced but turned through a right angle (Fig. 53 (*c*)) and, in Fig. 53 (*d*), with the fringes spread out to give a contrast effect described later (p. 90).

Another example of the use of two-beam fringes is given in Fig. 54,

Fig. 53. Effects using a two-beam interferometer. The specimen is electropolished α-brass, lightly strained, original ×480, enlarged ×2.

(*a*) Monochromatic light, fringes widely spaced.
(*b*) Exactly the same setting as for (*a*), but using white light.
(*c*) Monochromatic light, fringes closer spaced than in (*a*) and oriented in a perpendicular direction.
(*d*) White light fringes spread out to give a contrast effect very similar to phase contrast (compare Fig. 29 of Chapter 5).

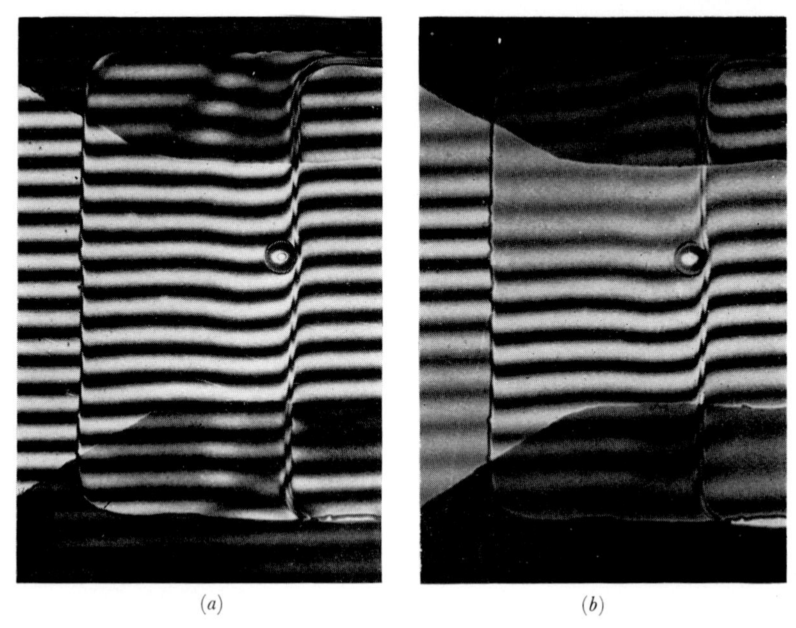

(*a*) (*b*)

Fig. 54. Two-beam fringes used to measure the thickness of evaporated layers of metal in an integrated electrical device.
(*a*) Monochromatic and (*b*) white light.

where the thickness of an evaporated layer is being measured. The illustrations in the next section on multiple-beam interference fringes might also have been produced with two beams to show the use of this type of interferometer.

Multiple-beam Interference

It is possible to obtain interference fringes from opaque objects using much simpler equipment than that required for two-beam fringes; moreover, this technique (multiple-beam interferometry) is capable of giving very much greater sensitivity than two beams. On the other hand, it demands critical conditions and is not readily capable of allowing adjustment of fringe spacing and orientation.

The essence of the technique is that conditions are arranged for the incident beam in Fig. 45 (*b*) to be reflected many times between the optical flat OF (equivalent to P_1 in Fig. 45 (*a*)) and the specimen M (equivalent to P_2), part of the beam emerging parallel to the original reflection at each of these new reflections. A whole family of emergent beams is thus generated and, with the correct conditions, these interfere and produce fringes exactly duplicating the spatial distribution of two-beam fringes but having a different intensity distribution. Instead of following a cosine squared curve, as found for two-beam fringes, multiple-beam fringes follow a more complex distribution of intensity. This has the effect, shown in Fig. 55 (compare Fig. 46) of making the dark fringes very much narrower and, at the same time, sharper. It is thus possible to estimate the height of a feature very accurately and to pick out very small features. Under the best conditions, not easily attained in metallography, it is possible to measure fringe displacements of 1/100 of a fringe spacing, i.e. of 1/100 × 0·27 μm = 27 Å or 10 atoms.

Although the conditions for production of multiple-beam fringes have been known for many years, it is only during the last twenty years, principally through the work of Tolansky, that these conditions have been translated into a practical system of operation. Tolansky has described the theory and application of multiple-beam techniques in his

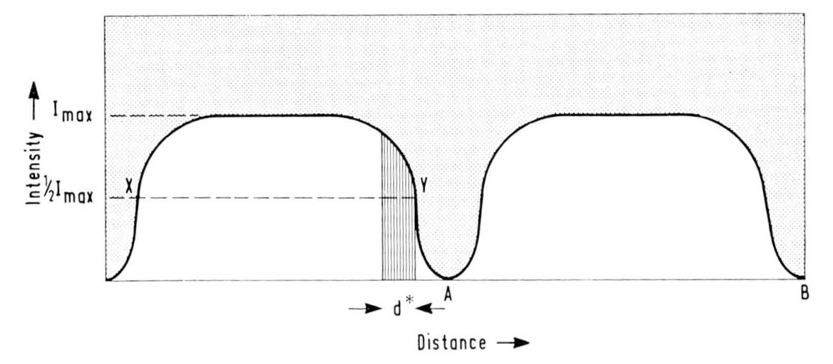

Fig. 55. Intensity distribution for multiple-beam fringes, approximately as found with typical metallic specimens. Measurements to 1/5 of d* (d* = 1/20 of λ/2 from Fig. 46) are possible.

book[2], and we will not repeat the theory here but concentrate upon describing the attainment of the correct conditions. Wherever possible, however, some indication of the theoretical reason behind each condition will be given.

CONDITIONS FOR MULTIPLE-BEAM INTERFEROMETRY

Specimen Reflectivity: The first requirement for multiple-beam interferometry is to maintain a sufficient intensity of the reflected beams. Several features are concerned in this. Perhaps the most important is that the specimen itself should have a high reflectivity; fortunately, many polished metals do reflect sufficient light for some dozen or fifteen reflections to be obtained. This will give fringes considerably sharper than those obtained using two beams. However, to obtain the full sensitivity of multiple-beam techniques as many as 80 or 90 reflections are required and this demands a reflection of 90 per cent of the light at the specimen surface. Again, fortunately, there is a simple way of arranging this—by evaporating an opaque layer of silver on to the specimen *in vacuo*. The equipment for doing so is now a standard adjunct to the electron microscope, and the preparation of silvered specimens is therefore possible in many laboratories. Increasing the reflectivity of the specimen by this means would not be of much use unless the surface contours were precisely preserved. Tolansky has been able to show that this is so to almost atomic dimensions.

Reference Plate: This has to meet four conditions: it must be optically flat, thin enough to go between the objective and the specimen, have material of high reflectivity on its upper surface, and this material must also have a low absorption, so that the light is not attenuated unduly by repeated passage through the reflecting material. Once again, it happens that vacuum-deposited silver has the required properties of high reflectance and low absorption. The plate itself may be an optical flat, specially prepared to be very thin, but for any but the most critical applications a selected microscope cover slip will serve. It will be found that many slips from an average boxful will be very close to being optically flat. The best of them can be selected by placing random pairs face to face and viewing them in a monochromatic light, or even in light which has strong peaks in its distribution of wavelengths, such as that from the common fluorescent tube. Fringes will be seen in the plane of the cover slips; if they are both flat these will be straight and number only three to five across their diameter. A few trials will soon eliminate the unsuitable slips, and show which of the rest are closest to being optically flat.

The thickness of silver required to give the required reflectivity may be measured with appropriate densitometers, but is conveniently gauged by trial and error, actually observing which silvered slips produce the sharpest and most contrasty fringes with a particular specimen. As a first approximation, however, the colour of the silver film viewed with transmitted light is an excellent guide; a film of 90 per cent reflectivity has a characteristic purple hue. It can readily be arranged to view the film in transmission whilst it is being deposited in the vacuum equipment, and in this way the effect of variations in the time or amount of evaporation may be followed.

A disadvantage of silver films is that they oxidize fairly rapidly, sometimes within a few days, and then produce spotty images and, more importantly, begin to absorb more light. Good protection from oxidation may be obtained by evaporating a layer of silicon dioxide on to the silver. For the majority of metallographic applications, an evaporated aluminium film has sufficiently low absorption, and it lasts almost indefinitely. Certain dielectric materials, such as are used in blooming lenses, may also be used to make reflecting layers. It is possible to build up multi-layer combinations which have low absorption together with high reflectivity for one wavelength, but high transmission for another. This permits viewing of the normal image without removal of the cover-slip reflector.

Monochromatic Light: It will be clear from an examination of the interference equation (1) that the position of a fringe is shifted laterally as wavelength changes, each wavelength picking out a particular wedge separation. To exploit any interference technique to the full it is therefore necessary to restrict the wavelength of the light to as narrow a band as possible.

With unsilvered metal specimens, it is often difficult to obtain sufficient intensity in the fringes for photography, so the most intense source available should then be used.

Collimation: The incident beam must be closely parallel; clearly, divergence of the beam will spread the reflected beams widely after multiple reflections, and so the conditions for obtaining multiple reflections from the area of interest will be destroyed. On a metallurgical microscope it seems adequate if the illuminating beam is cut down by closing the aperture stop, after having set the microscope to give Köhler illumination. However, if the separation between the specimen and the reference plate is on the large side, as is unavoidable sometimes on deformed specimens, the fringe sharpness can be improved by stopping down further by inserting an artificial aperture. The amount of light reaching the photographic plate in consequence may be very small and exposures of over half an hour required. Of course, this is a circumstance which would normally be avoided, but it may sometimes be worth the trouble; an example of this would be during tests which are difficult to repeat, such as long-term creep tests.

Normal Incidence of Illumination: It is also important that the illuminating beam should be incident normally to the area of the specimen being studied. This is necessary in order to prevent the multiple reflections spreading over too wide an area of the specimen, and is one of the principal reasons dictating the use of objectives of relatively low magnification. In metallurgical work a 16 mm objective (approximately $\times$ 15) is usually the most powerful that can be employed in multiple-beam interferometry.

It is of considerable assistance in obtaining fringes if the departure of the incident beam from the normal can be controlled. Some older microscopes had their beam-splitting mirrors mounted on a draw tube; this could be moved laterally or rotated and adjusted until the fringes were as sharp as possible. (This procedure was used in obtaining the photographs for Figs. 56–58.) Most modern microscopes, with their tendency to monolithic alignment and push-button operation, are without this flexibility of control over the illumination; consequently the tilting must then be imposed on the system comprising the specimen and reference plate. Holders capable of sensitive adjustment of the tilt of the specimen and the reference plate are available; it must be emphasized that the degree of sensitivity required is great, for it is dictated by the fact that fractions of a wavelength of light are significant.

Bearing the above conditions in mind, the practical procedure for obtaining multiple-beam fringes from a metal specimen may be summarized as follows. Firstly, obtain as close contact as possible between the specimen and the "silvered" cover slip; focus in the normal way with the aperture stop opened to the usual maximum. Then, with monochromatic light, cut down the aperture to its minimum and possibly refocus very slightly to bring the fringes into view. If, however, no fringes

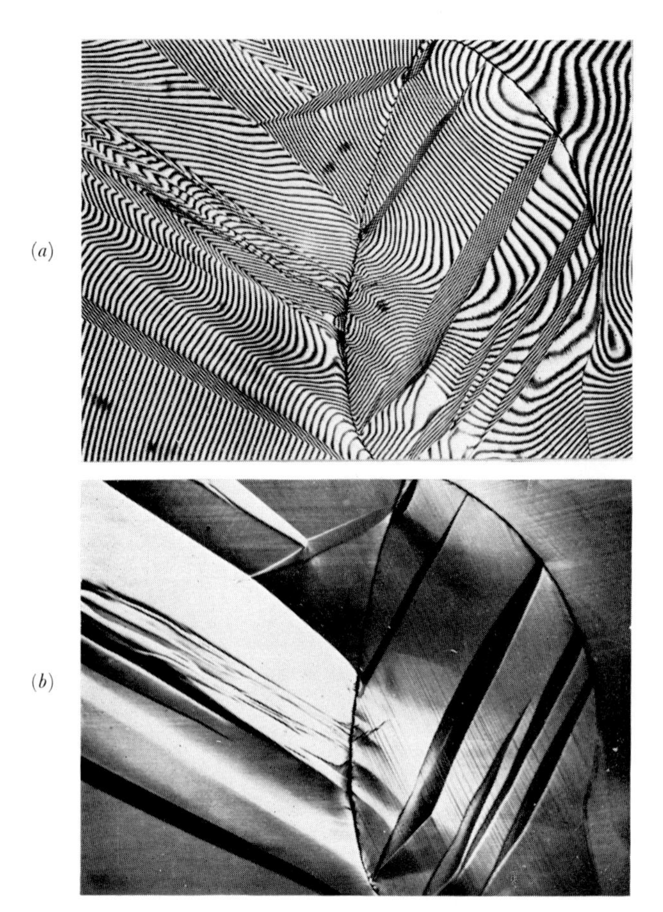

Fig. 56. Twins in deformed zinc, showing that slopes of twins in each grain are parallel and also showing accommodation kinking. Electro-polished, unetched. ×200.
(*a*) Multiple-beam interferogram.
(*b*) The same area with oblique illumination.

are seen, it may be necessary to improve the normal incidence of the beam by tilting the illuminator or the specimen. These adjustments are then "played" until a satisfactory sharpness and contrast of the fringe is obtained. If the fringes are even then faint, it may be that the separation between the specimen and the reference plate is too great, or perhaps that the plate is the wrong way round, with the silvered side away from the specimen. Another possible cause of faint fringes is mismatch between the reflectivity of the specimen and the reference plate.

Light Sources for Interferometry

Monochromatic light may be obtained in a number of ways. Usually, a wavelength in the green part of the spectrum is chosen, but the Watson interferometer is supplied with a sodium lamp (yellow light). White light

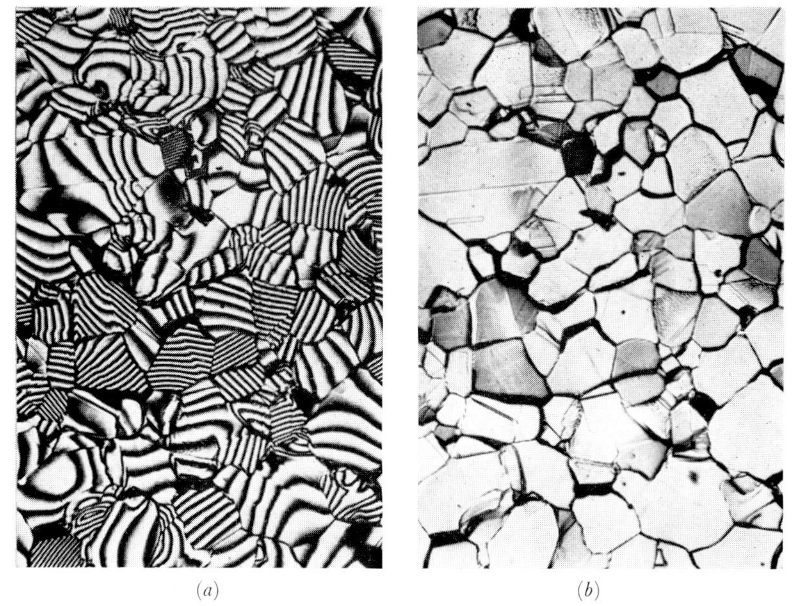

(a) (b)

Fig. 57. Tilting of grains with little internal deformation (the fringes remain straight and parallel) on a creep specimen of a lead-thallium alloy; originally chemically polished to give a flat, featureless surface; unetched. × 150.

(a) Multiple-beam interferogram.
(b) Bright-field illumination.

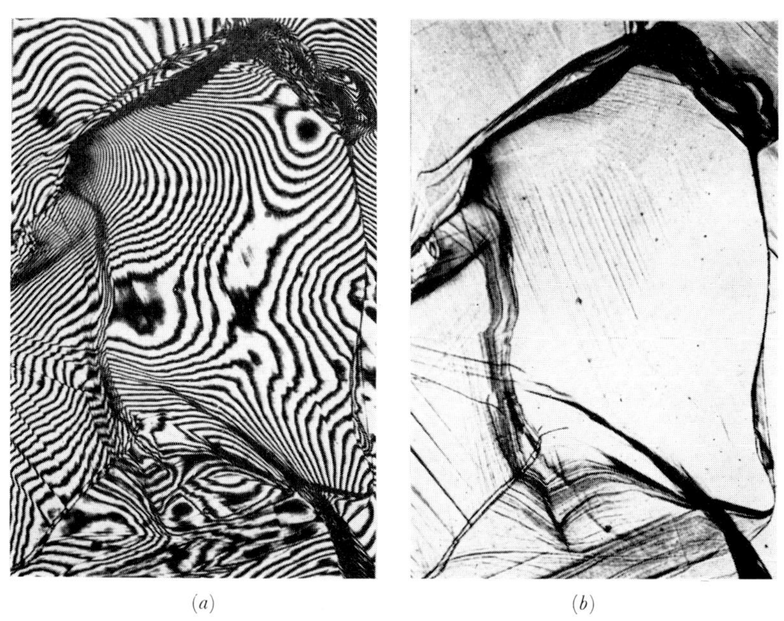

(a) (b)

Fig. 58. Grain-boundary sliding and migration, triple-point folding, fine and coarse slip, formed during the creep of lead; chemically polished, unetched. × 150.

(a) Multiple-beam interferogram.
(b) Bright-field illumination.

can be filtered with a suitable glass, which passes only a narrow band of green light. This has the disadvantage that only a very small proportion of the generated intensity is made available for production of the fringes. A better procedure is to use a mercury arc lamp and a mercury-green, neodymium, glass filter to remove wavelengths other than that of 0·54 μm. A low-pressure mercury lamp, such as is used in street lighting, but with its outer glass envelope removed, gives a greater intensity at this wavelength than does the more expensive high-pressure lamp. However, the latter is sometimes a standard fitting on microscopes. A thallium arc lamp gives intense monochromatic light ($\lambda = 0{\cdot}54$ μm) and is standard equipment on one Linnik-type instrument.

Applications

Several typical applications of two-beam and of multiple-beam interferometry have been given during the preceding pages. Some other examples of the use of multiple-beam interferometry are given in Figs. 56 to 58, which also compare this technique with other ways of examining the same areas. It will be obvious that the kind of information given by the interferometry is particularly valuable in studies of deformation or of surface finish. It has also proved valuable in the study of the surface tilts produced by martensitic phase transformations. The principal limitation, apart from any difficulties in obtaining fringes at all, lies in the relatively low lateral magnifications which can be used. On the other hand, interferomatic techniques give precise quantitative data very much more readily than, say, phase-contrast or stop-contrast illumination; this is an important advantage.

Interference Contrast

So far, the interference techniques described have been those which have been specifically arranged to give fringes which form a contour map of the specimen. It is possible to change the appearance of fringes and obtain images very similar to those produced by phase-contrast illumination; that is, the interference technique is used to give enhancement of contrast between features of interest and their neighbouring backgrounds.

The simplest method is that of broadening the fringes until only one or two fringes cover the area in which detail resides. The fringes then tend to take up the shape of some features, as in the example shown in Fig. 53 (*d*). The mechanism of achieving this is simple in theory, for it merely involves bringing the reference mirror into a position which is almost parallel to the general background level of the specimen. With multiple-beam interferometry of metals this may present difficulties, because the distance between the specimen surface and the reference mirror is critically small and the angle between them set by the most prominent asperities of the specimen; variation of this angle is not therefore very easy to bring about. With two-beam interferometers the necessary adjustments may be possible through the variations in optical path and the tilting of the mirror which are built into the instruments. Failing this, it is not difficult to arrange some method of tilting the specimen itself to obtain the desired contrast.

Another large family of techniques for obtaining interference contrast is based on the use of birefringent plates and polarized light. At least two of these methods have been developed to the stage where they are available commercially as attachments to the polarizing microscope. They will be described here, since the images are closely related to the normal interference-contrast images, but the details of terminology on polarization should be sought in Chapter 7. These techniques tend to be rather specialized and the theory on which they are based is complex. Some general description will be given here. For those desiring more detail and rigour, a very complete description of the various kinds of interference contrast and of the theory of the phenomenon is given in Françon's book[3] (pp. 105–128).

POLARIZING INTERFERENCE

The basic principle of all polarizing-interference microscopes is image duplication in some birefringent material. The complete image consists of two parts, one formed by the ordinary ray and a second formed by the extraordinary ray. We can, following Françon, represent these images by lines with humps upon them, the humps indicating a change (in phase) imposed on the plane incident wavefront by the variation in level of the specimen, which is otherwise featureless—that is, uniformly reflecting.

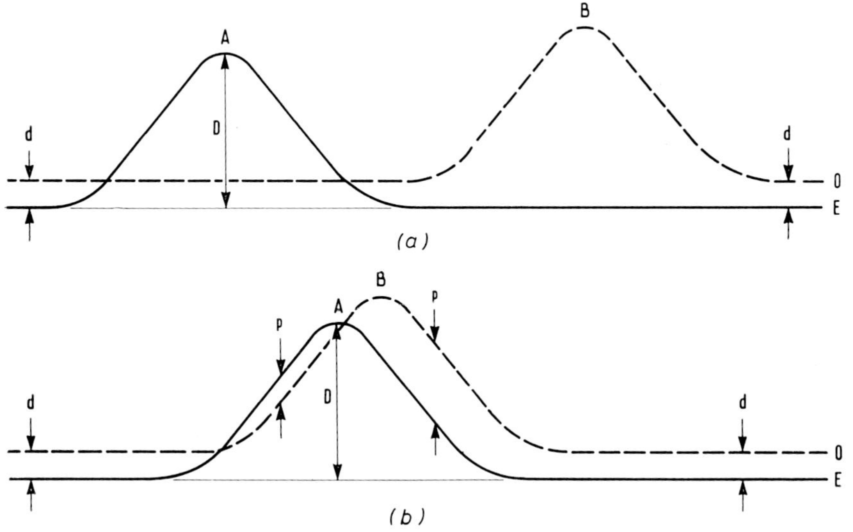

Fig. 59. Schematic representation of polarizing interference. A plane wavefront is reflected from a specimen having variations in level and then duplicated by birefringence to give O and E images (see text).
(*a*) With complete image separation.
(*b*) With image overlap (differential type).
(After Françon.)

This is indicated in Fig. 59 (*a*), where the path difference between the general level and that of the feature examined is *D*, the height of the humps. The birefringent plate causes image duplication, represented in the diagram by the ordinary image O and the extraordinary image E. In general, between these there will be a shift of the hump laterally as well as of the whole curve in the phase direction, resulting in a phase difference *d* at the places representing the uniform background level. Thus the invisible feature is made visible by suitable manipulation of the image at A and B, where the respective path differences are now $D - d$ and $D + d$.

The background level of the specimen will appear as a purple hue when the image is viewed in white light with crossed polarizer and analyser, this being the first-order colour for interference with polarized light. The path differences $D - d$ and $D + d$ give rise to other colours, according to their values, and so the feature becomes visible as a series of coloured fringes related to the structure (in terms of different levels). If we can adjust the components to make $d = 0$ the whole field is made dark except at A and B, which remain coloured with hues appropriate to the phase change *D*. When *d* is very small and *D* is also not too great, the image becomes rendered almost entirely in terms of dark grey with lighter areas at A and B, and this image is then very similar to a phase-contrast image. Should the lateral extent of the features be too great, the images A and B overlap, the image becoming confused: this may often happen on metallurgical specimens which are, in general, very complex and involve many large features as well as very small ones.

If the two images are only shifted by a small amount they will overlap as in Fig. 59 (*b*) but yet have the phase difference *d* as before. Then the feature will be made visible by the polarization-interference effects contingent upon the differences in path *d* and *p* on the background level at points as indicated. The difference between *p* and *d* will be less great then than in the previous case involving the path difference *D*, but nevertheless, it will give rise to contrast effects; because the contrast varies according to the "slope" of the hump, this method is called a differential one.

FRANÇON'S INTERFERENCE EYEPIECE

This method is based on the differential method just outlined, and uses a Savart plate to produce the image duplication.

A Savart plate consists of two pieces of quartz cut 45° from the optical axis and crossed. This results, as shown in Fig. 60 (*a*), in the following relationships for a ray of particular wavelength incident upon the first plate L. It splits into the ordinary ray O_1 and the extraordinary ray E_1 which then pass through the second plate N. This in turn splits the rays so that the ray O_1 becomes the extraordinary ray E_2 and the ray E_1 becomes the ordinary ray O_2. Thus the emergent ray E_1O_2 is in the plane of the paper, but O_1E_2 is parallel to it but in a plane above (or below) the plane of the paper. The reason for this is that the axis of N is not in the plane of the paper but at 45° to it in the figure. The effect, therefore, is not only to displace E_1O_2 to the right, as shown in Fig. 60 (*a*), but also to displace O_1E_2 an identical distance perpendicular to the plane of the paper, as symbolized by the dotted line. The relative positions of the various rays are also shown in Fig. 60 (*b*).

Where the incident ray is normal to L, the emergent rays E_1O_2 and O_1E_2 are in phase, because their paths are equal and symmetrically

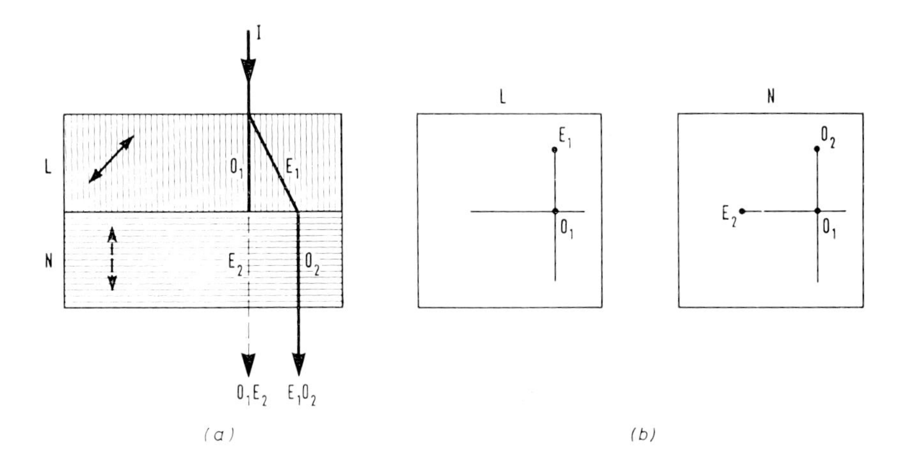

Fig. 60. The Savart plate.
(*a*) Showing paths of O and E rays in the two quartz plates L and N.
(*b*) The relative positions of rays in L and N.
(After Françon.)

placed with respect to the incident ray. If the plate is tilted about an axis normal to the plane of the figure, the paths differ for the two rays, according to the amount of tilt. If the illuminating beam is limited by a slit and this is imaged on the Savart plate, it can be made to select a particular fringe from the coloured series seen in uniform white light. In this way, tilting the Savart plate moves the image into different fringes and so controls the coloration and contrast of the final image.

In practice the arrangement is very simple. The Savart plate (Sa) is incorporated in an ocular E (Fig. 61), and the specimen is illuminated using the slit D to limit the beam. The light from the objective passes through a polarizer (pl) before reaching the ocular and an analyser (al) is placed between the ocular and the eye.

The intensity of illumination is rather low because light is limited by the slit and diminished by the use of crossed Nicols; the slit also imposes restrictions both on the resolution and on the evenness of illumination. Nevertheless, the device is flexible and useful, particularly in the absence of more complex equipment.

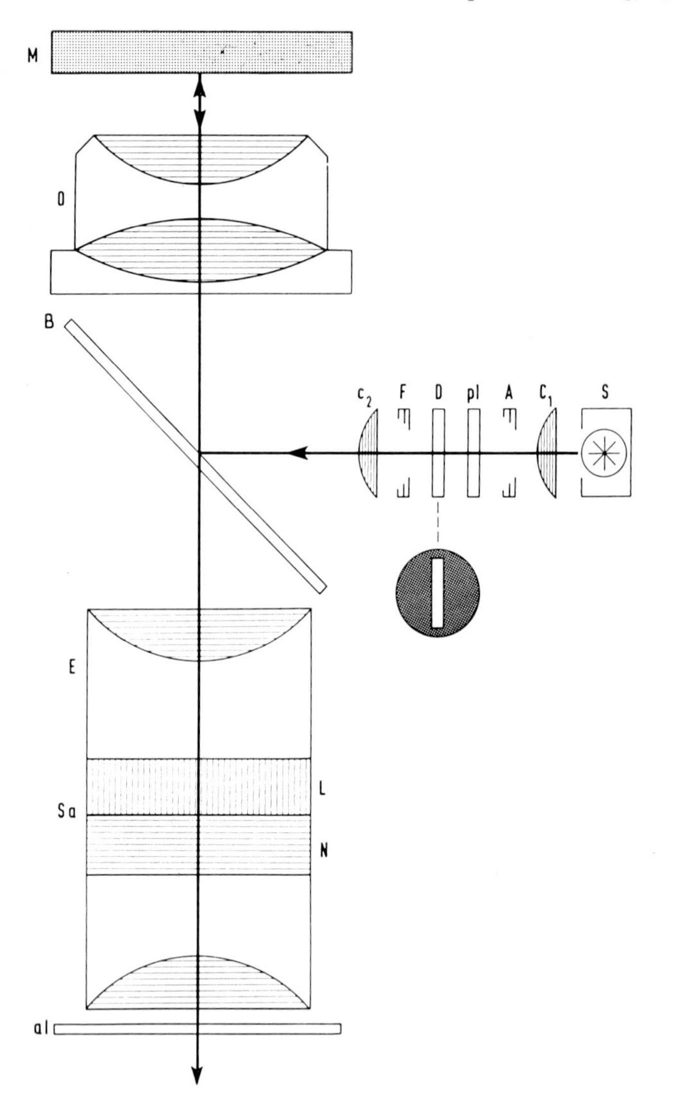

Fig. 61. Françon's (polarizing) interference eyepiece; a Savart plate (Fig. 60) is incorporated in the eyepiece and illumination is through a slit D.

NOMARSKI'S INTERFERENCE DEVICE

This also uses the differential method and gives images very similar to those produced using Françon's eyepiece, but it is not subject to the limitations imposed by the slit illuminator. The birefringent unit is, in this case, a Wollaston prism. This consists (Fig. 62) of two quartz plates W_1 and W_2 cut at an angle and in such a manner that the optic axes are at right angles to each other; that of W_1 is normal to the plane of the diagram, that of W_2 is in the plane of the diagram and at right angles to the optic axis of the microscope. The Wollaston prism is incorporated at the focal plane of the objective O and can be moved across the optic axis as indicated in the diagram.

The incident light is plane-polarized after passing through the polarizer (pl) and the critical setting of the illuminating system ensures that the beam is focused at I in W_1. After reflection at the specimen it is then

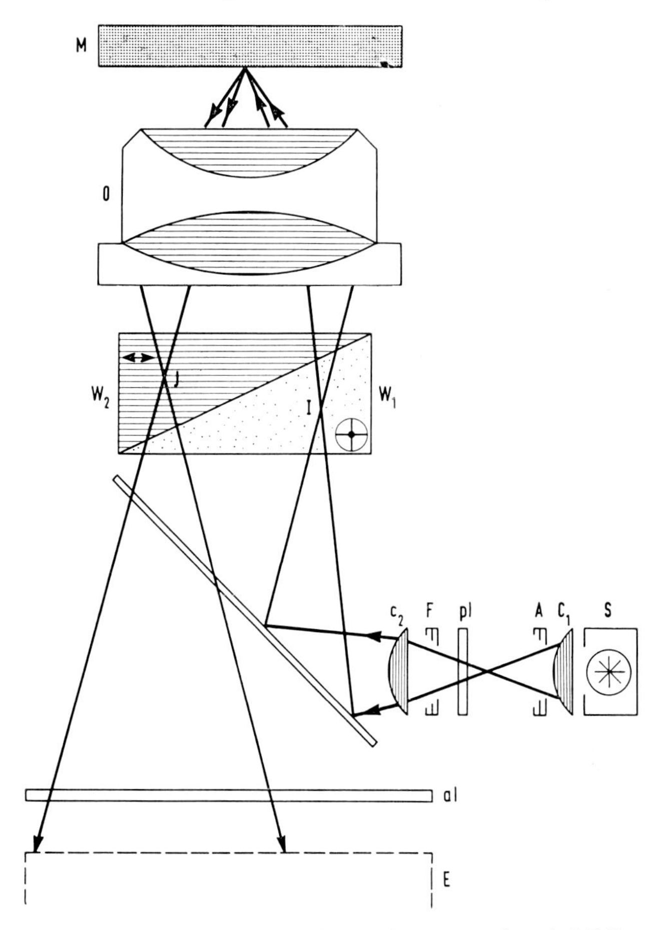

Fig. 62. Nomarski's (polarizing) interference optics. A Wollaston prism W_1W_2 is incorporated close to the back of the objective. The device gives a range of contrast effects according to the setting of the Wollaston prism.

brought to focus at J in W_2. The geometry of the prism is such that I and J are symmetrical, and the path difference of the ordinary and extra-ordinary rays in their passage to the specimen equals that produced on their return through the prism, but is of opposite sign. With the analyser (al) in the crossed position, the equalization of the path differences of all O and E rays leads to their suppression by the analyser; a darkened field results. In terms of our previous discussion of the differential method, the path difference D has been made zero. However, any displacement of the rays by reflection at tilted areas of the specimen leads to their return through a non-symmetrical position in W_2. In this way, the path difference

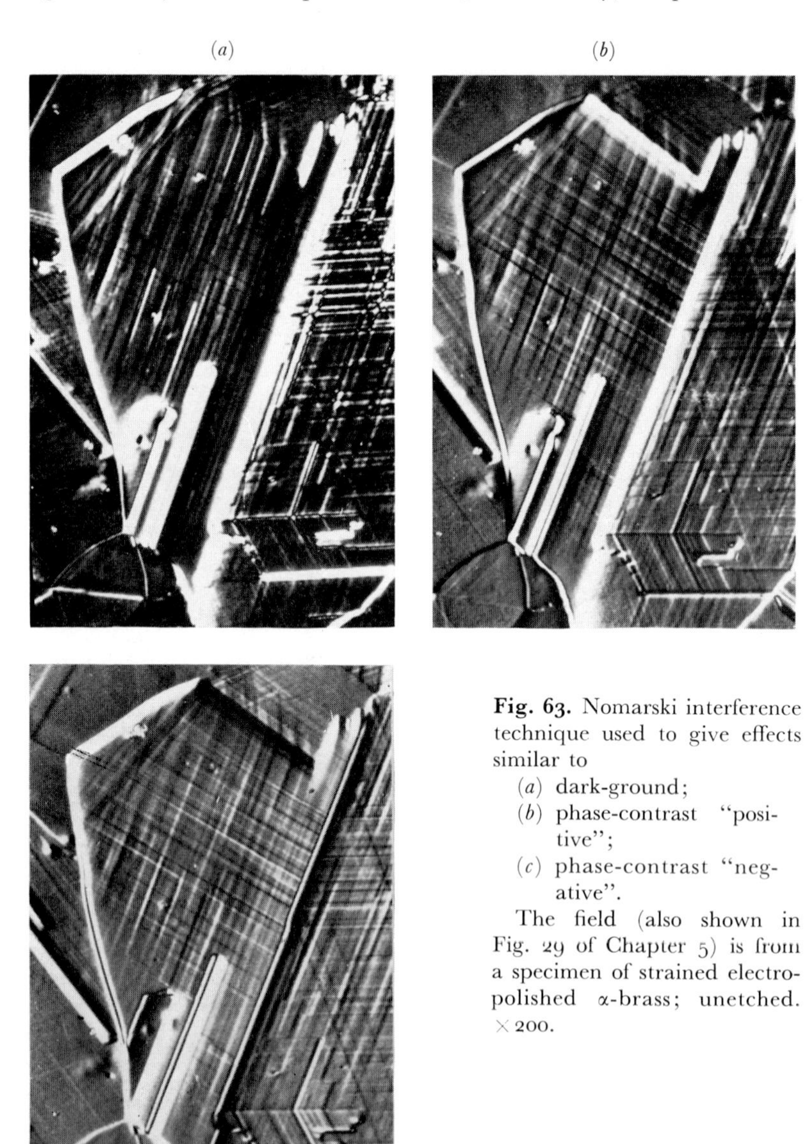

Fig. 63. Nomarski interference technique used to give effects similar to
(*a*) dark-ground;
(*b*) phase-contrast "positive";
(*c*) phase-contrast "negative".

The field (also shown in Fig. 29 of Chapter 5) is from a specimen of strained electropolished α-brass; unetched. ×200.

d between the O and E rays is made to be proportional to the degree of tilting, and the analyser now passes a proportion of the recombined rays. The result is to render the tilted area with an intensity dependent upon d.

The field therefore has an appearance very similar to that obtained with dark-ground illumination, having tilted features bright upon the dark background. Fig. 63 (*a*) is an example of the use of the Nomarski technique to give this kind of contrast. Other settings give results similar to positive or negative phase-contrast or to oblique illumination; examples are given in Figs 63 (*b*) and (*c*).

By moving the relative positions of the two prisms, D can be set at values other than zero. The background then takes on a hue dependent upon this path difference, because the analyser now passes the whole spectrum except for a wavelength equal to D. For example, when $D = 0{\cdot}565\ \mu m$ (a wavelength in the yellow region), the background becomes white light minus yellow and therefore of a purple hue. Tilted areas impose additional path differences as before, and give rise to other hues.

When D is set to give a grey-green background, the device is capable of yielding high contrast in the image; it then appears to be somewhat better than conventional phase-contrast. A comparison is shown in Fig. 64.

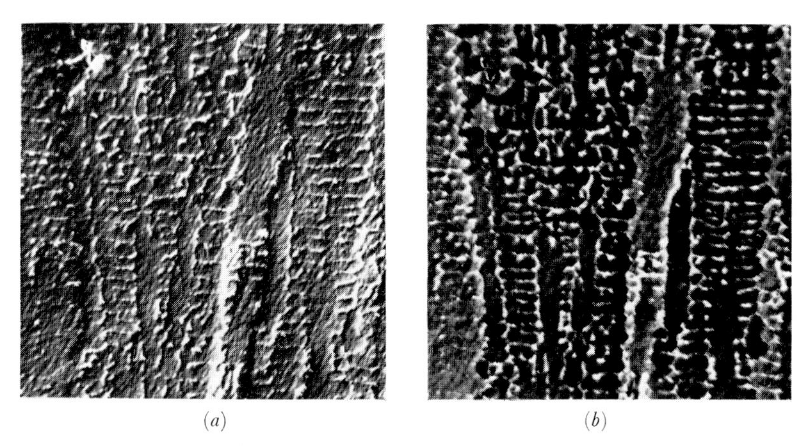

(*a*) (*b*)

Fig. 64. (*a*) Nomarski interference compared with (*b*) conventional phase-contrast (negative).

The specimen is cast copper, mechanically polished but unetched. Slight relief has rendered the dendritic coring visible with the special illumination. × 100.

The settings which result in coloured images are also useful because of the sensitivity of most eyes to changes in colour; the structure becomes "etched with light" in a pleasing and informative way.

Miscellaneous Applications of Interference Techniques

It is possible to obtain some idea of surface topography using a two-beam interference to produce fringes of low contrast by the following method. Some Canada balsam is dissolved in xylol and a drop of solution placed on the area of interest and allowed to evaporate. A thin film of balsam

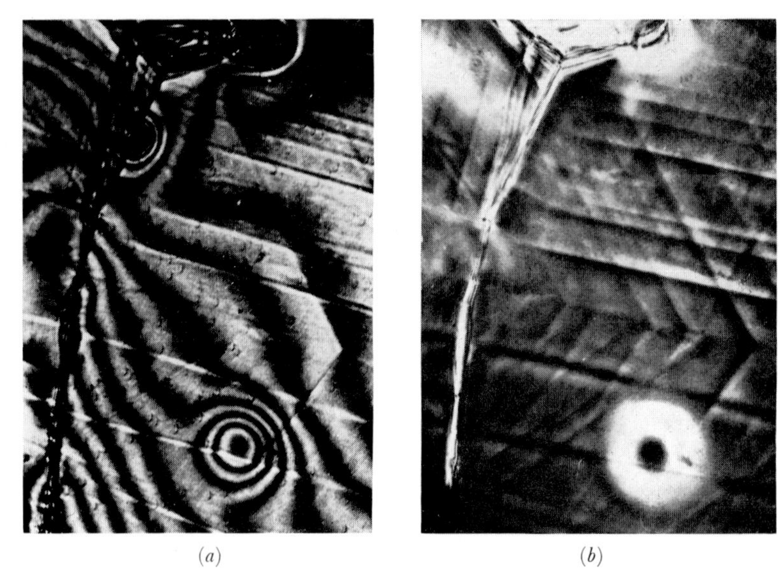

(*a*) (*b*)

Fig. 65. (*a*) Two-beam fringes produced by the use as a reference plate of a thin deposited layer of Canada balsam on the specimen. (The area shown is identical with that in Fig. 32 of Chapter 5.)
(*b*) Same area, positive phase-contrast. Deformed lead, unetched. ×200.

remains and it is then possible to evince two-beam interference fringes from two reflections, one at the upper surface of the balsam, which is approximately flat, and the other at the specimen surface. Fig. 65 shows such fringes. They are of very low contrast, but the occasion does arise when this method will give some information which may be difficult to

Fig. 66. Two-beam fringes from a plastic replica of a worn extrusion thread, inaccessible to examination with a microscope. The replica was "silvered" to enhance fringe contrast. This interferogram was used to decide how much machining was required to restore the surface. Unetched. ×480, enlarged ×2.

obtain in any other way. Presumably the contrast would be improved by evaporating sufficient silver on to the balsam layer to give closer matching of the intensities of the two beams.

Sometimes the area of interest may be inaccessible to standard interferometers, because the specimen is too heavy or because the area is too far below the general surface. It may then be possible to replicate the area to be studied (see Chapter 9) and obtain two-beam or multiple-beam interference fringes from the replica. This procedure has been used successfully to study scoring on the surface of a helical thread used in a large extrusion press for cable sheathing. The two-beam fringes were not of high contrast but sufficient to determine the amount of grinding required to repolish the grooves and also to draw useful conclusions concerning the particles causing the score marks. Again it was possible to enhance the contrast (Fig. 66) by silvering the replica; in this case a relatively thick, opaque layer of silver was required.

The Light-cut Microscope

We include a description of the light-cut microscope in this chapter, not because it is an interference microscope, but because it does give essentially the same topographical information as the interference method.

The method was originally described by Schmaltz[4], whose name is sometimes given to microscopes embodying the principle. A fine band or

Fig. 67. Showing the principle of the light-cut and light-profile microscopes. Shadows of power lines, obliquely cast on to the corner and window of a building, reveal its stepped shape as a "contour" which can be used quantitatively.

wafer of light is shone obliquely on to the surface of the specimen and the pattern it makes is then viewed obliquely from the other side. An analogue to the system may be seen in shafts of light falling across a flight of steps, or, as in Fig. 67, the shadow of power cables on the corner of a building.

There are two distinct versions of the light-cut microscope, the first being embodied in commercial instruments designed for engineering studies of surface machining, and the second being used more specifically for metallographic work. In the former instruments the magnification is limited vertically to a relatively low of × 400. The incident light is

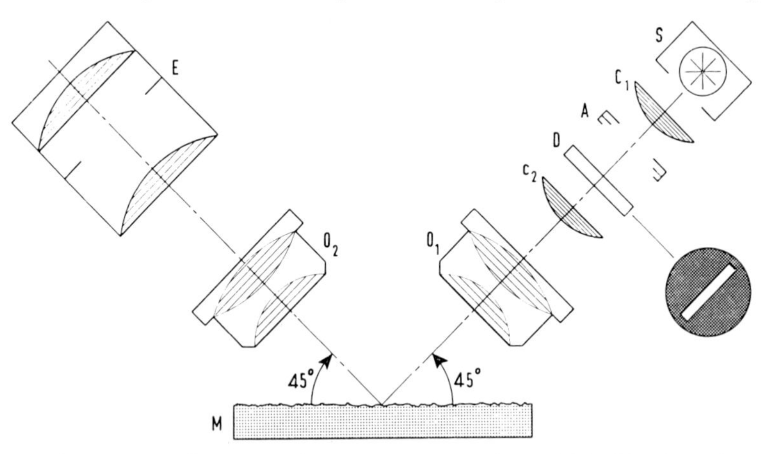

Fig. 68. The light-cut microscope; the specimen M is illuminated obliquely through the slit D and the objective O_1 and viewed obliquely from the other side using objective O_2,

limited by a fine slit D and focused on the specimen by an objective O_1 (Fig. 68), which is at 45° to the plane of the specimen. The image is formed by a second objective O_2, also at 45° to the surface and at right angles to O_1. The image gives a profile of the surface with a magnification of $\sqrt{2}$ times the lateral magnification M. Fig. 69 shows the kind of image

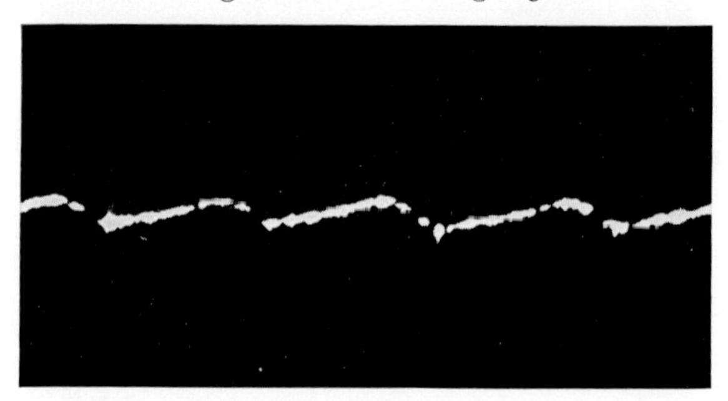

Fig. 69. Light-cut image of machining grooves on aluminium; *vertical* magnification, × 150 approx.

formed; it will be seen that the profile has a characteristic "chopped" appearance, which may make interpretation of the detail a little difficult. The rest of the specimen is not imaged and this sometimes makes identification of a particular feature uncertain; however, with semi-regular features, such as machining grooves, neither of these is necessarily a disadvantage.

Many metallographs may be adapted to the use of the light-cut method, as described by Tolansky[5]. The essential conditions are:

(i) oblique illumination. This can be obtained very conveniently on at least one commercial instrument (the Vickers' microscope) using the metal-tongue illuminator. This consists of a 60° sector of polished metal which replaces the beam-splitter;

(ii) a high-power, immersion objective, to obtain oblique "viewing" of the obliquely cast beam; and

(iii) monochromatic light. In this case the magnification in depth M_1 is given by

$$M_1 = (2M/n) \tan \theta$$

where M is the lateral magnification of the objective, n the refractive index of the immersion medium and θ the effective angle of incidence. It is more convenient to determine M_1 experimentally than to try and compute it and this is done by obtaining the profile of some known step—the edge of a shim or something similar. Alternatively, a groove or step may be measured interferometrically and used to calibrate the microscope.

The disadvantages given for the light-profile technique may be overcome if the image of a fine wire is substituted for that of the fine slit. Tolansky,

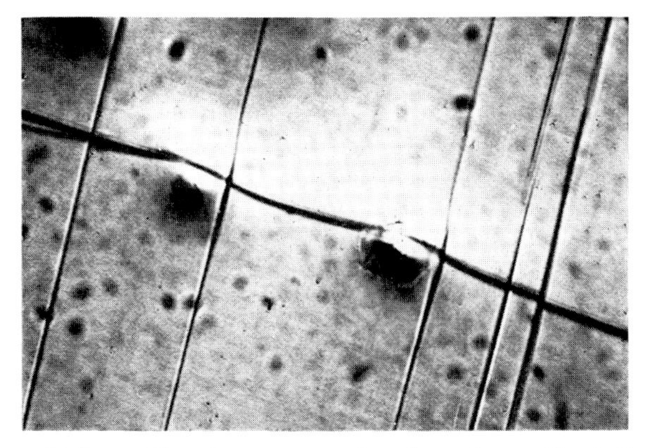

Fig. 70. Light-profile microscope used to show the profile of slip steps and a polishing pit on lead; lateral magnification ×350, vertical magnification ×400 approx.

who also introduced this variant, called it the light-profile microscope. The wire, which should be less than 10^{-4} in. diameter, is mounted at the field stop of the microscope and the conditions (i)–(iii) met as described above. Fine wire of suitable diameter is commonly used in resistance strain gauges. It may be affixed to a brass washer with glue and mounted in a holder which pushes into the surround of the field stop. The image now consists of the usual, but somewhat oblique, bright-field one, with the shadow of the wire, moulded to the profile, superimposed on it. Fig. 70 is an example. A series of parallel wires or even two orthogonal sets may be built up if the gain in information per field of view seems warranted.

The technique is capable of giving large lateral magnifications at the same time as magnification in depth, although the latter are smaller than those obtained with interferometry. It is also a quantitative method and, moreover, an inexpensive one. It deserves more attention than appears to have been given it by metallographers. A commercial instrument using the light-profile principle is available.

REFERENCES

1 W. E. Hoare and B. Chalmers, *J. Iron & Steel Inst.*, 1935, **132,** 135.
2 S. Tolansky, *Multiple-beam Interferometry of Surfaces and Films*, Oxford, 1948 (Oxford University Press).
3 M. Françon, *Progress in Microscopy*, Oxford, 1961 (Pergamon).
4 G. Schmaltz, *Technische Oberflächenkunde*, Berlin, 1936 (Springer).
5 S. Tolansky in *Properties of Metallic Surfaces*, London, 1953 (Inst. of Metals).

7 The Polarizing Microscope

An application of polarized light has already been described in the chapter on interference techniques, where polarization was used as a means of obtaining a special kind of two-beam interference. In this chapter we shall begin with some notes on the nature, production and properties of polarized light and then describe applications to metallography. It will not be possible to do more than give details of a few of the more widely used applications and then indicate other uses, because the field is very large. Fortunately, a comprehensive book by Conn and Bradshaw[1] is available for those who wish to pursue this topic further. The principal reason for the fullness of exploitation of polarized light in reflected-light microscopy is that it has been a most important tool in the hands of mineralogists, who are often concerned with anisotropic crystals; as will emerge in the discussion of polarized light, it is of particular use with such materials.

Polarized Light

The vibrations in a light beam are in directions normal to the direction of propagation of the light, but their direction varies too rapidly to be detected; light may therefore be considered to have a random direction of vibration—see Fig. 71. Under some conditions of reflection, and on passing through some crystalline materials, some of the directions of vibration are suppressed and there may be rotation of others, so that specific directions of vibrations are much favoured. In the limit, only

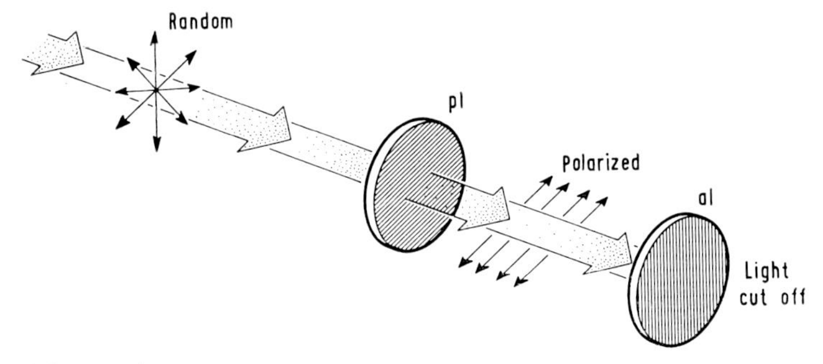

Fig. 71. As normally produced, a beam of light has random directions of vibration (left); after passing through the polarizer (pl), the vibrations are confined to one plane, so that the beam is cut off by the analyser (al) in the "crossed" position.

vibrations in one plane may remain after the reflection or refraction. The beam is then said to be *plane-polarized*, that is, the vibrations are confined to a particular direction between specific poles.

Light is an electromagnetic vibration and it is in the electrical interactions of light waves with matter that polarization arises. A rigorous and detailed examination of polarization in these terms is beyond the scope of this book but, as in other chapters, an attempt will be made to give some idea of the physical processes behind the observed phenomena. The vibrations spoken of are, according to Maxwell's theory of electromagnetic waves, the electrical components of the light waves, so that in a plane-polarized wave it is the electric displacement which is confined to a particular plane.

The most usual way of producing a plane-polarized beam of light is to pass light from a normal source through a transparent plate of some specific, *optically active*, material. This unit is called a *polarizer*. There are several kinds of material which may be employed as polarizers, their principles of operation all being somewhat similar but having important differences. One of the most common polarizers used to be a Nicol prism. More often today the polarizer is made from a synthetic material produced in sheet form and called by the proprietary name of Polaroid. As these two kinds of polarizers represent classes of material acting according to different principles their properties will be examined in more detail.

Birefringence and Dichroism

Calcite is a *birefringent* crystal. When cut in a particular manner with respect to the crystal symmetry, it has the property of producing a double image of an object viewed through it. This property is a result of an incident ray of light (Fig. 72) being doubly refracted to form two rays within the crystal, the ordinary ray O, which obeys the normal law of refraction, and the extraordinary ray E, which has a different law of refraction, because it travels with a different velocity to the O ray. Along a specific direction in the crystal, called the *optic axis*, the O and E rays move with the same velocity, but in other directions they differ, so that starting from a particular point P the wavefront of the O ray in Fig. 72 is a circle and that of the E ray an ellipse whose minor axis equals the diameter of the circle.

The physical events leading to this result are complex, and depend upon the particular kind of anisotropy found in crystals such as calcite. Some idea of the nature of the phenomenon may be conveyed by the following general description. The molecules of the calcite crystal may be viewed as coupled units (dipoles) which are set into resonant vibration with the electric displacements of the light wave. The coupled units in calcite are such that the resonant vibration has two orthogonal components, one of which is dominant. Further, the molecular units build into a structure having, in this context, a screw-like character in one specific direction; thus the interaction with the electric vector gradually forces the latter to assume preferred directions consonant with the two vibrational modes of the molecules. In this way the incident ray becomes split or polarized into the O and E vibrations. It is found that the plane of polarization of, say, the O ray, gradually rotates as the thickness of the crystal is increased; this is to be expected from what has just been said.

It is also found experimentally, and can be deduced from the detailed theory, that the phase of the E ray is some angle ϕ different from that of

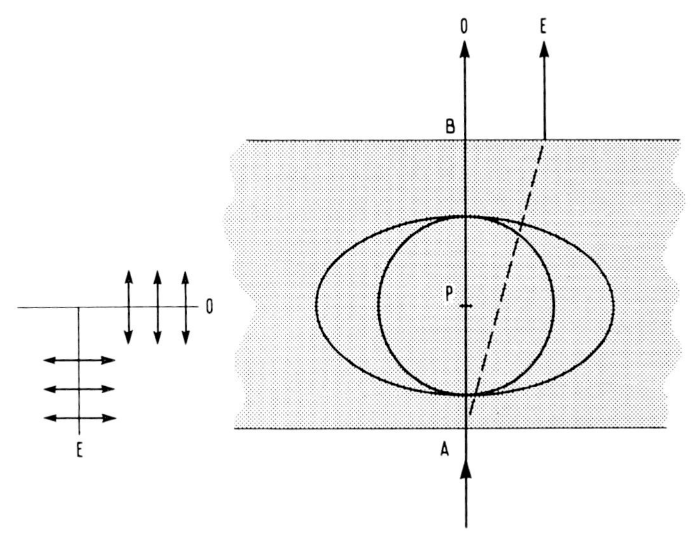

Fig. 72. Wavefronts of the ordinary ray O and the extraordinary ray E in a block of birefringent crystal. The optical axis of the crystal is along AB. The arrows at the side of the diagram indicate the appropriate directions of polarization for the O and E rays. The wavefronts of the two rays are represented by a circle and an ellipse.

the O ray. This is a consequence of the mode of vibration of the molecules, the dominant vibration being out of phase by this angle with respect to the minor one.

To use calcite in the production of a plane-polarized wave it is necessary to suppress the E ray. This could be done by having a sufficiently thick crystal, so that the divergence of the E ray from the O ray is large enough to be able to mask it off. It is more convenient to achieve this by cutting two prisms from a calcite single crystal, as in Fig. 73, and cementing them together with Canada balsam to form a composite prism known as a Nicol prism. The cuts are made so that the optical axes are as indicated in the figure and the angle of the interface is such that the O ray is totally reflected at the internal (balsam) interface. In this way the emergent E ray is isolated from the O ray and is plane-polarized, its plane of vibration defining the plane of polarization of the Nicol.

The less expensive way of producing plane-polarized light employs Polaroid sheet. Originally, a Polaroid sheet consisted of crystals of a specific material, aligned within a transparent mounting medium, which caused the O and E rays to be separated. The crystals in this case are not only birefringent but also have the additional property that the absorption is very much greater for the E ray than for the O ray. Such crystals are called *dichroic*. This property may be used to lead to the suppression of

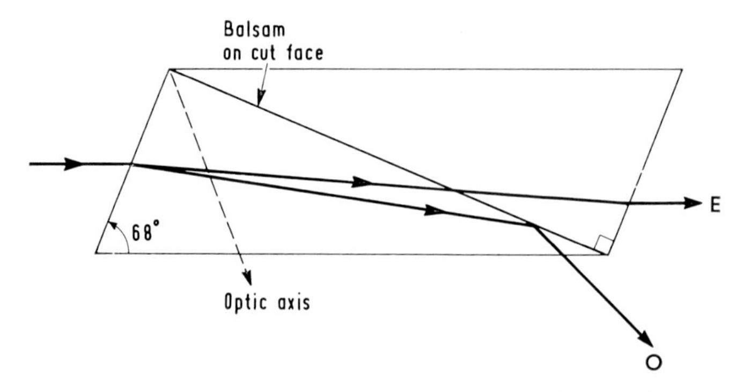

Fig. 73. A Nicol prism—two pieces of a calcite single crystal are cut and cemented together as shown, the optic axis being as indicated. The O ray is reflected at the balsam layer, thus isolating a plane-polarized E ray.

the E ray and production of a plane-polarized O ray. More recent Polaroid sheets do not contain oriented crystals but are made instead from plastics which have long molecules; these can be aligned by mechanically stretching the sheet; if necessary, some addition of a chemical dye is made to give the absorption required for dichroism.

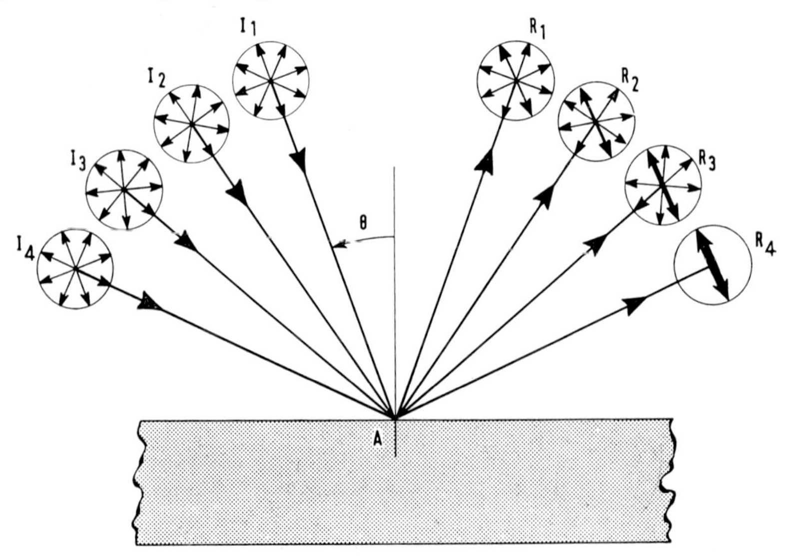

Fig. 74. Polarization on reflection (Brewster's Law). As the angle of incidence θ of rays I_1, I_2, etc. is increased the unpolarized incident ray is reflected at R_1, R_2, etc. with an increasing component plane-polarized as indicated by the thick arrows; at a critical angle, a maximum of polarization is reached, defined by Brewster's Law.

If the plane-polarized light transmitted by a polarizer is examined after passing through a second Nicol prism (or sheet of Polaroid), which can be rotated about the direction of propagation of the beam of light, it is clear that there will be a position of the second birefringent plate which cuts off the light completely (Fig. 74). This position will be attained when the plane of polarization of the second Nicol is at right angles to that of the first. When the two plates are in this position they are said to be *crossed* or to be set to give *extinction*, and the second plate is called an *analyser* to distinguish it from the polarizer. This arrangement finds widespread application, for it is a means of measuring the effect of changes imposed on a plane-polarized beam by material inserted between polarizer and analyser. For example, if this material rotates the plane of polarization, it will be found that the analyser has to be turned through a corresponding angle before the light is again cut off completely. In this way the *extinction position* or angle of extinction may be measured.

What has been said so far has carried the implicit assumption that the light considered is monochromatic. With white light the situation is usually further complicated by the fact that the birefringent or dichroic properties depend upon wavelength. As will be discussed later, this can be used to advantage in obtaining colour contrast in various circumstances.

Polarization on Reflection

Polarization occurs on reflection from most substances, even when they are isotropic. For anisotropic materials the effect depends in part on the principles we have already discussed for refraction by anisotropic media. However, a characteristic of reflection at any angle other than normal incidence is the inducement of some degree of polarization in the reflected beam.

Suppose an incident beam I falls at an angle θ to the normal upon a sheet of glass, as shown in Fig. 74. Even when the incident beam I is completely without any polarization it will be found that the reflected beam R contains a component which is plane-polarized. This may be ascertained quite simply by passing it through a Nicol prism or a piece of Polaroid. As θ is increased the proportion of polarization will rise from zero at normal incidence to a maximum at some value θ_c and then decrease again. This critical angle or *polarization angle* is defined by *Brewster's Law:*

$$\tan \theta_c = n_2/n_1$$

where n_1 and n_2 are respectively the refractive indices of the medium above the reflecting surface and of the reflecting medium itself. For glass in air θ_c is about 57°.

The plane of polarization of the reflected beam (i.e. the plane to which the electric displacement of the light vibration is normal) is the plane of incidence. This is defined by the incident and reflected rays and the normal to the surface at the point of reflection—IAR in Fig. 74.

The physical model for this kind of polarization is again based on the idea of oscillating atomic dipoles; the atoms are nuclei with their "bound" electrons. Thus the effect is not only found with dielectric materials such as glass or liquids, where the only electrons are in the bound state, but it is also found with metals where "free" electrons exist. In the present context, the importance of these free electrons is that they are responsible for the high reflectivities of metals compared with dielectric substances.

Returning to the discussion of Brewster's Law, we note that the polarization angle corresponds to the condition in which the maximum number of dipoles is vibrating along the direction of the reflected beam, inducing a coupled vibration (i.e. polarization) in this beam. In an isotropic medium this turns out to depend only upon the angle of incidence, because the isotropy allows atoms to be paired to form approximately equivalent couples whatever the orientation of the crystal; of course, in non-crystalline materials such as glass, the isotropy arises from a random arrangement of molecules rather than from symmetry, as in crystals.

On the other hand, with an anisotropic metal, reflection is affected by the vibration of dipoles which, by their departures from a symmetrical arrangement, override the influence of the angle of incidence. This leads to polarization of the reflected beam, but in a manner that is specifically related to the orientation of the crystal considered.

Elliptical and Circular Polarization

Should the incident beam be partially or completely plane-polarized in any but the plane of incidence, the state of vibration of the reflected beam is further complicated. This is because it now contains both the initial and induced polarized vibrations. It is convenient to compound these vibrations—that is, to imagine a particle to be forced to move under the influence of both of them and so describe some appropriate locus, usually non-linear. It can readily be shown that the resultant motion is, in general, an ellipse. Suppose that the two vibrations are along orthogonal axes, of unequal amplitudes and different in phase by $\pi/2$. Then, as indicated in Fig. 75 (*a*), the compounded vibration is an ellipse and the light is said to be elliptically polarized. When the amplitudes of the two vibrations are equal, the ellipse degenerates into a circle, giving circularly polarized light, Fig. 75 (*b*). More general results of the same kind may be obtained with vibrations which are not orthogonal; the axes of the ellipse are then turned in relation to those shown in Fig. 75. A further case is possible; this is when the vibrations are in phase or 180° out of phase. There is then further degeneration of the ellipse to a linear vibration across the diagonal of the rectangle in which the ellipse in Fig. 75 is inscribed. This could be thought of as rotation of the original plane of polarization.

When elliptical polarization is introduced into a plane-polarized beam, for example by reflection at an oblique angle, a Nicol in the crossed position no longer extinguishes all the light. An amount determined by the nature of the introduced vibration is transmitted by the analyser. It is this particular circumstance which is especially exploited in the examination of metals using polarized light. However, as will be indicated in the

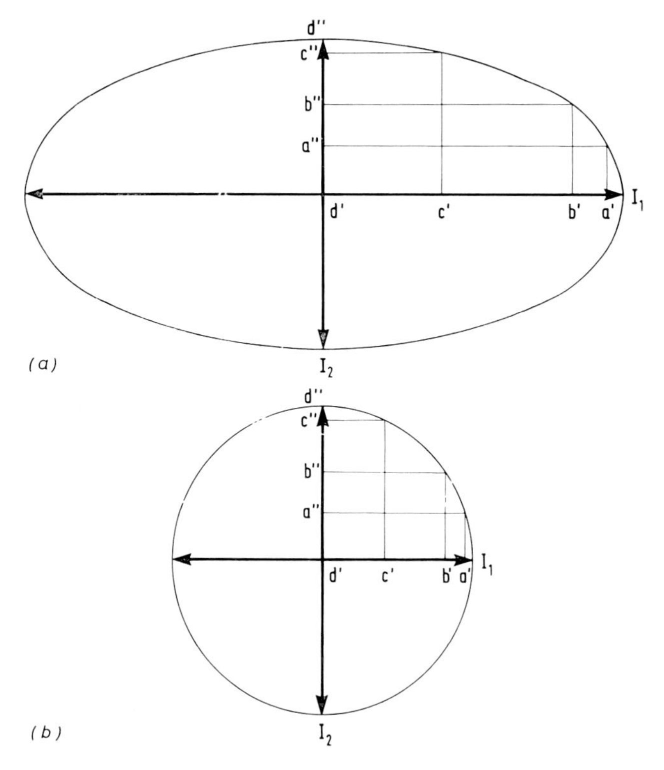

Fig. 75. Combining plane-polarized vibrations I_1 and I_2 as (*a*) elliptically, and (*b*) circularly polarized light.
These result according to the phases and amplitude of I_1 and I_2, as indicated.

next few pages, there are other ways in which the use of polarized light may assist in the study of metals.

Transparent Inclusions

Many inclusions in metals are transparent and their appearance when examined between crossed Nicols may be characterized both by the reflection from the upper interface and by reflections from internal faces, when the transmission of light through the inclusions is involved. If examined in polarized white light, transparent inclusions may show characteristic colours. The reflected light is extinguished by crossing the polarizers, but the transmitted light is dependent upon the optical properties of the inclusion, and a characteristic range of wavelengths is seen, because these are polarized in a plane which is passed by the analyser. If light of certain wavelengths is absorbed preferentially by the inclusion, its colour will change as the stage holding the specimen is rotated. This phenomenon is known as *pleochroism*. Yet another effect may be obtained from the regularity or lack of regularity of the lower surface of the inclusion. Polarized light refracted into the inclusion will be reflected at this lower surface and, according to its angle of incidence, will be more or less converted to light which is no longer plane-polarized. After passing through the analyser light and dark fringes or crosses will be seen as a result.

Equipment and Setting

POLARIZERS, VERTICAL ILLUMINATOR

It has been indicated previously in Chapter 4 that the polarizer is placed somewhere between the aperture stop and the beam-splitter. It is very desirable, if not absolutely necessary, that a plane-glass beam-splitter be used. Any prism, sector reflector, or device which gives an unsymmetrical beam tends to give uneven illumination and some change in polarization across the field, because of the varying angles of incidence of the beam on the specimen.

There is, of course, variation in the angle of incidence of rays reflected on to the specimen from a plane-glass illuminator, but these are symmetrically disposed and have a minimum adverse effect.

A more important result of using a plane-glass illuminator is the variation in angles of incidence of rays reflected from the illuminator itself. In accordance with the discussion leading up to Brewster's Law, a plane-polarized beam will be made more or less elliptical according to the position of rays and their angles of incidence on the reflector. Thus, undesirable *depolarization*, as it is sometimes called, may be minimized if the plane of polarization is arranged to be parallel (or perpendicular) to the plane of incidence. The Foster prism, which is available only on Bausch and Lomb microscopes, automatically overcomes these difficulties and dispenses with the necessity for use of a separate polarizer.

It is an advantage to use a coated plane-glass illuminator with polarized light; indeed, this is so for all metallography, and coated illuminators are now a standard fitting on the majority of metallographs. The coatings increase the reflectivity of the upper surface by employing some material of high refractive index such as zinc sulphide, and reduce strong reflections from the lower surface by "blooming".

The analyser may be placed anywhere between the vertical illuminator and the eye. For much qualitative work it can be held by a cap which fits over the ocular. When angles of extinction are required, the analyser should be in a holder capable of rotation about the optic axis of the microscope and preferably have a pointer moving over a scale marked in degrees. A convenient place for mounting the analyser is often just after the vertical illuminator (e.g. see Fig. 13, page 20).

OBJECTIVES

With proper adjustment, any good quality objective may be used with polarized light. The principal defects are likely to be variations in uniformity of polarization resulting from strain within the glass (or fluorite) components of the lens. Strain causes glass to become birefringent by destroying its isotropy locally and this leads to ellipticity in a plane-polarized beam. Clearly, this could confuse the message from the specimen examined. Specially selected, strain-free, objectives may be obtained for work with polarized light. They should be handled with particular care, for even finger pressure is capable of introducing transient strain which is readily detected by polarized light. For qualitative work, and the less exacting kinds of quantitative applications, standard objectives may be sufficiently free from strain. If possible, lenses should be selected when they are purchased on the basis of a minimum strain. This may be assessed in the following way. A flat, cleanly polished piece of an isotropic metal is used as a specimen and the objective focused in the usual way. The polarizers are then put into the crossed position and the ocular removed. A characteristic fringe pattern, known as an *isogyr*, will then be seen—see Fig. 76 (*a*). It takes the form of a cross, whose arms denote the

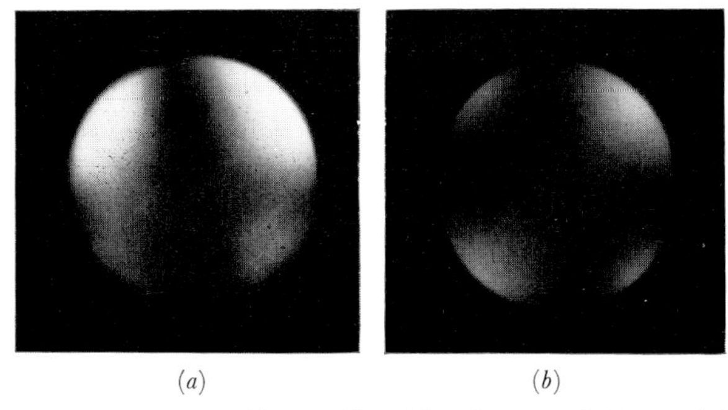

(*a*) (*b*)

Fig. 76. Fringe pattern ("isogyr") arising from strain in an objective lens and revealed by polarized light; the dark areas of extinction correspond to the planes of polarization of the polarizer and analyser and vary in extent as the lens is rotated—compare (*a*) and (*b*). In using the lens with polarized light, it is rotated to obtain the maximum area of extinction.

vibration directions of the polarizer and analyser. Strain will distort this pattern so that it changes as the lens is rotated in its mount—see Fig. 76 (*b*). Besides remaining little affected by rotation, a good lens will have a broad centre to the cross, so that a useful portion of the field does, in fact, show extinction when the polarizers are crossed.

A best setting should be found for a lens in which there is strain by rotating it until the area of extinction, judged from the appearance of the isogyr, is at a maximum.

N.A. has an effect on the uniformity of polarization, because the peripheral rays from an objective of high N.A. are incident upon the specimen, with sufficient obliqueness to introduce noticeable ellipticity on reflection.

Thus it is desirable to use the lowest N.A. compatible with the magnification and resolution desired. Alternatively, with a given N.A., which therefore fixes the magnification, the aperture stop may be reduced to increase the contrast, remembering that this procedure is inimical to the highest resolution. It is possible, at some further sacrifice of resolution, to limit the area of the lens in use to the two directions corresponding to the vibrations in the polarizer and analyser. This is done by inserting a special stop, or *epiphragm*, into the illuminating train; the stop has a cross-shaped opening, corresponding to the isogyr pattern.

ILLUMINATION

The intensity with polarized light is often low, so the brightest source available is an advantage. The only other restriction is that a source having a reasonably uniform spread through the spectrum is desirable when coloured effects are being used.

SETTING FOR EXTINCTION

For many purposes the crossed position of the analyser, which gives extinction, may be attained by straightforward visual examination. Many analysers are mounted on rotating arms marked in degrees and it will be found difficult to distinguish between positions over a few degrees close to the true extinction. Often this is of no consequence. Should a more accurate and reproducible setting be required, this may be done using a photoelectric cell mounted instead of the ocular on the microscope. A minimum reading of the output from the cell then corresponds to extinction.

ACCESSORIES

A number of aids to the interpretation of effects obtained with polarized light are available and are commonly used in mineralographic work. These consist of pieces of birefringent material such as quartz cut in various specific ways and mounted so that they can be placed somewhere between the crossed polarizers. In metallography the most useful of these is known as the *sensitive-tint*, *magenta-tint* or *whole-wave* plate.

A sensitive-tint plate consists of a slice of some birefringent material, usually quartz, cut parallel to the optic axis of the crystal; it is of such a thickness that the phase change between the emergent ordinary and extraordinary rays for green light ($\lambda_g = 5400$ Å) is equivalent to λ_g. Suppose the microscope to be focused on a plane isotropic specimen, with the polarizers crossed, and the sensitive-tint plate interposed between them with its optic axis at 45° to their vibration directions. The plate will introduce no ellipticity into plane-polarized light of wavelength λ_g, because the ordinary and extraordinary rays emerge from it in phase ($\lambda_g \equiv 2\pi$) recombining to reform the original plane-polarized beam, as indicated in Fig. 77. The analyser will therefore continue to extinguish light of wavelength λ_g. All other wavelengths will pass through the analyser to some extent, the amount depending upon the degree of ellipticity introduced for each wavelength by its passage through the plate; this in turn is a consequence of the phase difference arising between the ordinary and extraordinary rays of each wavelength. With the set-up just described the light passed by the analyser will be white light minus the green wavelength; this is a magenta colour and hence one of the alternative

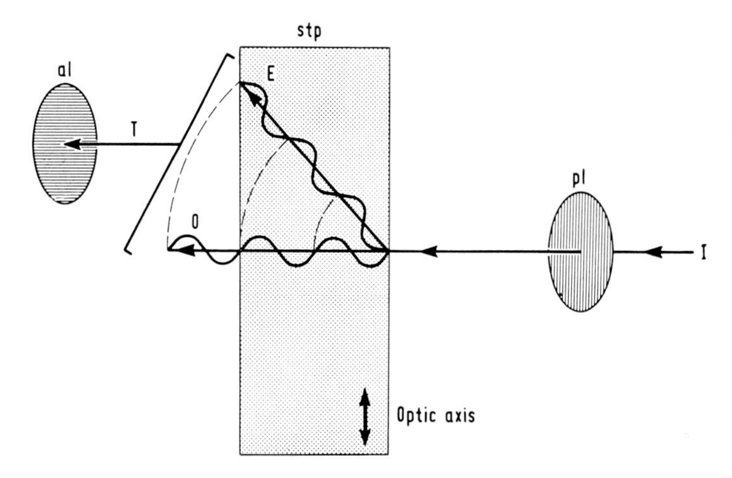

Fig. 77. Sensitive-tint plate. The incident ray I is plane-polarized; on traversing the plate (stp), which is of quartz, the O and E rays for a wavelength λ_g emerge the equivalent of λ_g out of phase; they recombine to form the plane-polarized ray T which is extinguished by the analyser (al). Other wavelengths are transmitted, because their E and O rays do not emerge λ apart. Hence, the analyser transmits white light minus green = magenta.

names of the plate. The reason for calling it a whole-wave plate is also obvious. The plate is used to introduce colour contrast into images obtained with polarized light; the method and an explanation of the effects obtained are given later (see p. 119).

Uses of Polarized Light in Metallography

ANISOTROPIC METALS

From what has been said, it follows that polarized light is a potential means of distinguishing orientation of anisotropic metals. With the polarizers set in the crossed position, grains of a polycrystalline aggregate should appear more or less bright according to the amount of ellipticity introduced by reflection, the ellipticity being related to the orientation of the various crystals. This is indeed found to be so. For grain contrast it may be said to be "etching with light". The contrast available depends intrinsically on the amount of anisotropy which the metal has, but the preparation of the polished surface may be of overriding importance. Strained or scratched surfaces, or oxide films, might suppress or falsify the true ellipticity arising from the anisotropy of the metal. Fig. 78 (*b*) shows a sample of zinc carefully prepared and so showing good contrast; in Fig. 78 (*a*) the contrast is degraded by poor preparation.

There may not be a great deal of advantage in the use of polarized light unless the metal is difficult to etch. It so happens that a number of metals

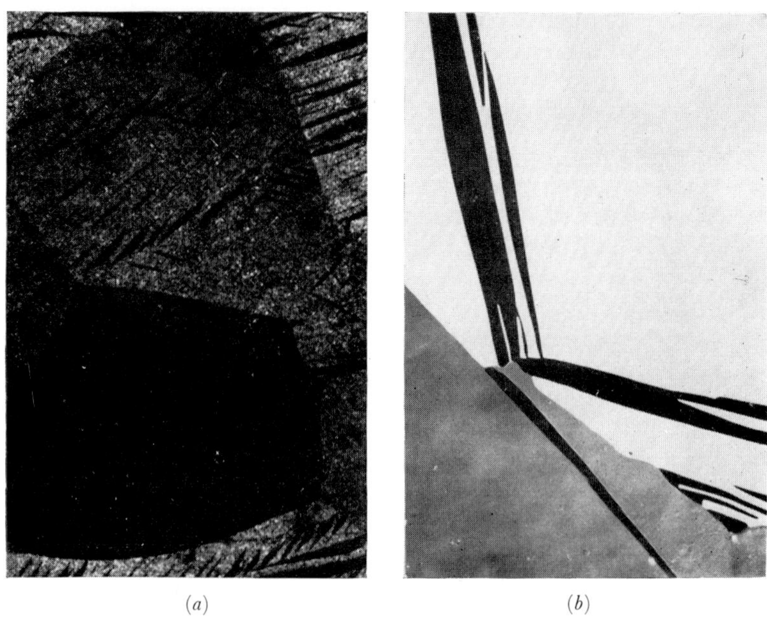

(*a*) (*b*)

Fig. 78. Effect of specimen preparation upon contrast obtained from an anisotropic metal (unetched) viewed with polarized light.
(*a*) Fine scratches and pits degrade the contrast.
(*b*) A good electropolish.
The specimens are of zinc. × 100.

Fig. 79. Polished but unetched uranium, viewed with polarized light. Grains, sub-grains, deformation twins and precipitates are seen. × 200. (C. J. Cocks.)

which have become of interest in the last decade or two are difficult to etch and are anisotropic; uranium, beryllium and zirconium are examples. For a more complete list see the Appendix to this chapter. Fig. 79 shows a photograph of polished but unetched uranium. A feature of grain contrast of this kind is that it can be controlled. For example, rotation of the specimen alters the relationship of the crystallographic directions of each grain to the vibration of the polarized beam, and particular grains then give extinctions at each setting of the specimen. (An illustration of this kind of effect is given in Fig. 83.)

ORIENTATION RELATIONSHIPS—PREFERRED ORIENTATION

Starting with one set of grains dark (i.e. extinction for these grains), the angle through which the stage has to be rotated to bring another set into extinction is a measure of the crystallographic relationship between the two sets. This information is insufficient to enable a unique description of the orientations to be made, because of the high degree of symmetry of metal crystals. However, if some additional simplifying fact is known, polarized light can be very useful in measuring misorientation. For example, if the misorientation is all within one grain, it may give a measure of the angular tilt between sub-grains or deformation bands.

The presence of a crystallographic texture may be detected using polarized light. The proportion of grains which show extinction together is some measure of this, but once again, unless some additional simplifying factor is present, the results can be confusing. The use of a sensitive-tint plate to give colour contrast facilitates this kind of work.

Polarized light has also been used to study recrystallization. The particular advantage here was that the orientation relationships between old and new grains could be followed at elevated temperatures without interrupting the test by cooling in order to etch.

ISOTROPIC METALS

Although, as we have seen, isotropic metals are inactive to polarized light, they can be treated to give grain contrast and reveal other orientation relationships. Two methods have been used, but it seems that three mechanisms might be involved in specific cases of the production of ellipticity.

ANODIC FILMS

One method, widely used with aluminium, is to grow a relatively thick oxide film electrolytically, by making the specimen the anode in a suitable electrolyte (see Appendix). Originally it was thought that the thickness of the oxide film varied according to the orientation of the underlying grain, and that birefringent properties of the oxide determined the ellipticity introduced on reflection. Perryman and Lack[2] showed this was not the case. They deposited, *in vacuo*, an opaque layer of silver on to the anodic film and found the grain contrast with polarized light to remain unchanged. Examination with the phase-contrast microscope then showed the surface of the oxide to be characteristically furrowed in each grain. It appeared, therefore, that the polarizing effect arose from double reflections, in a manner to be described below in connection with etched metals. Workers in France have attributed the polarizing effect to a more complex interaction with micropores in the oxide (Lacombe[3]). Anodic treatment has proved particularly useful in the detection of sub-grains formed during the creep of aluminium. Fig. 80 shows an area of an

Fig. 80. Bands of sub-grains in aluminium, deformed during creep, repolished and anodized, then examined with polarized light. × 100. (See also Fig. 27, Chapter 5.)

aluminium creep specimen where the banded nature of the sub-grains was clearly revealed using polarized light in this way.

An anodic treatment for examining magnesium has also been reported (Appendix C), but magnesium is sufficiently anisotropic to affect polarized light without such a treatment. Copper and its alloys, as well as a number of anisotropic metals, have also been anodically oxidized for work involving the use of polarized light.

THIN OXIDE AND SULPHIDE FILMS

Controlled oxidation or the formation of sulphides may also be used to produce thin surface films. These are often of such a thickness that interference colours are produced. However, the contrast is even stronger when polarized light is used and here the effect does seem to be one of birefringence within the oxide or sulphide film.

ETCHING

It is not too difficult to find specific reagents which can etch particular metals to form characteristic pits or furrows in each grain. These furrows are very similar to those found on anodic films. Ellipticity is introduced into plane-polarized light by these pits or furrows by double reflections from their sides. Each reflection (Fig. 81) is oblique with respect to these sides, even though the beam is normal to the general surface of the specimen. Accordingly, each reflection introduces ellipticity which is then

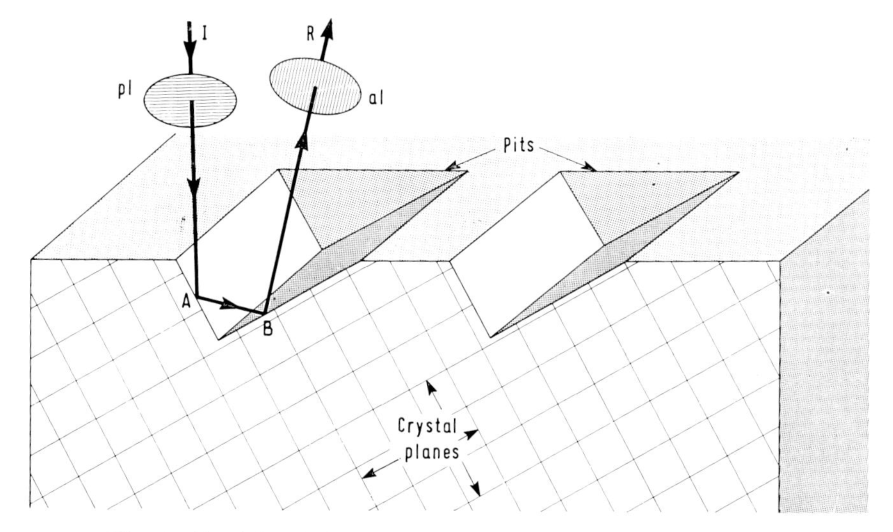

Fig. 81. Partial polarization of light by multiple oblique reflections at A and B on crystallographic pits or furrows on an isotropic crystal; the pits are formed using specific etching or anodizing techniques. A plane polarized beam so reflected is then partially passed by an analyser in the crossed position. The degree of polarization, and therefore image intensity, depends in this way upon the crystal orientation.

characteristic of the grain. This is because the furrows are made up of closely-spaced etch pits, the pits having faces of simple low-index plahes. Fig. 82 shows these furrows (compare Fig. 81) and Figs. 83 (*a*) and

Fig. 82. Furrows of the kind described in Fig. 81, produced on polished lead by a special etching reagent; negative phase-contrast. ×300.

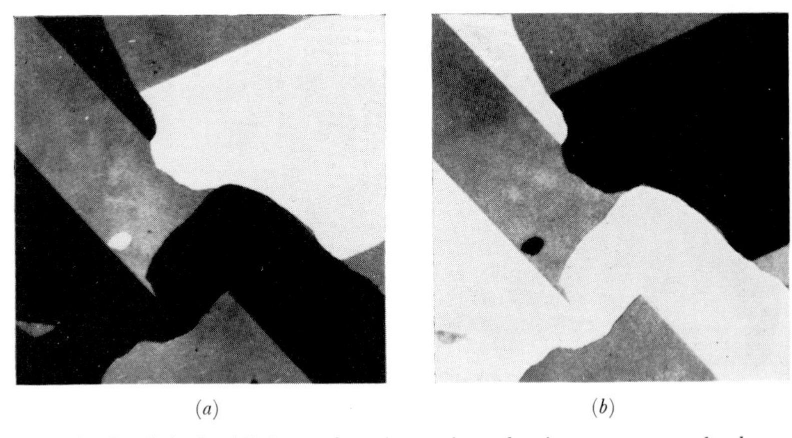

(*a*) (*b*)

Fig. 83. Polarized light used to give grain and twin contrast on a lead specimen etched as for Fig. 82. The contrast is varied by rotation of the specimen, compare (*a*) and (*b*). × 100.

83 (*b*) show an area of a lead specimen photographed with polarized light; the two photographs illustrate the effect of rotation of the specimen about the optic axis to control contrast. Fine pits of this kind are produced on α-brass on etching with one of a series of reagents; alcoholic ferric chloride is one of these which gives a very clean etch. Moreover, with this alloy, certain deformation traces are also revealed, as shown in Fig. 84.

Fig. 84. Polarized light used to give grain, slip line and twin contrast on α-brass deformed, electropolished and then etched (in alcoholic ferric chloride). × 100.

A similar effect from oblique reflections is obtained where surface relief is produced, for example, by a martensitic change or by polishing relief. Fig. 85 shows an area of a titanium specimen having a Widmanstätten structure brought into slight relief by mechanical polishing and viewed with polarized light. In this particular case some polarizing effect might also have resulted from anisotropy of the titanium.

Etch pits, cracks, folds, etc. may also be shown up as bright areas on a dark ground when isotropic specimens are view with polarized light.

Fig. 85. Widmanstätten structure in titanium, mechanically polished to produce slight relief; unetched and viewed with polarized light. ×200.
(Compare with Fig. 40, Chapter 5.)

USE OF SENSITIVE-TINT PLATE

The above contrast effects may be rendered in brilliant colour if the sensitive-tint plate is interposed between the polarizer and the analyser. Let us consider the colours obtained with a polycrystalline specimen of anodized aluminium. A grain which appears dark with crossed polarizers only will appear magenta with the tint plate in its 45° position, for it will give extinction for the green wavelength λ_g. Other grains will extinguish other wavelengths, according to the amount of ellipticity they introduce, and so each will show up with its characteristic colour (i.e. white light minus the wavelength suppressed). Rotation of the specimen causes the colours to change, for each setting of the stage picks out grains of a particular orientation to be those which give extinction for the green wavelength λ_g.

The introduction of colour contrast is not just a matter of obtaining a pleasing effect; it can yield additional information or make it easier to collect data accurately. Some degree of preferred orientation may be inferred if a proportion of the grains always change to the same colour on rotation of the specimen. Colour contrast also proves to be of assistance in the measurement of grain size. The eye finds it easier to register a change of colour from grain to grain than to identify the grain-boundary line when making measurements of lineal intercepts. Perhaps a more important advantage is that the colours enable identification of annealing twins to be made much more readily, and so twin boundaries are not erroneously counted as normal grain boundaries. The point here is that twin boundaries often go completely across a grain; the colours bring up the twin as a zone of one colour crossing another (the matrix). Even in the case of isolated twin boundaries, colour seems to assist in emphasizing the splitting of the grain into the two parts, matrix and twin.

OTHER USES

The uses to which polarized light has been put at times are numerous. The book by Conn and Bradshaw[1] should be consulted for details of these applications, especially the rather complex systematic methods which have been developed for identifying inclusions in steels, cast irons and aluminium alloys. Some examples of these structures are given in Figs. 86 and 87; these and similar structures are particularly beautiful

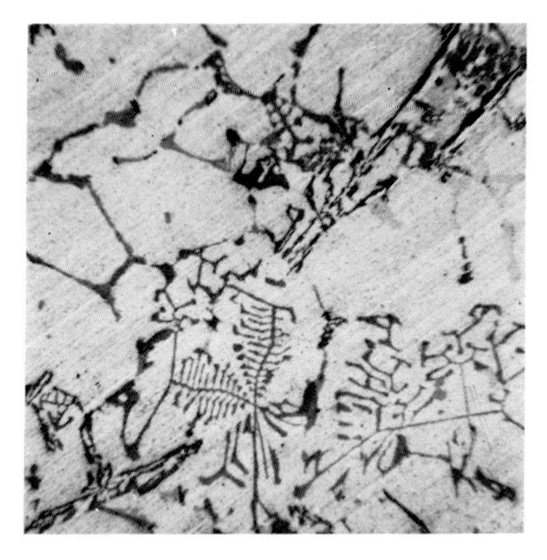

Fig. 86. Cast aluminium alloy (LM1) showing contrast between various types and orientations of intermetallics; the mechanical polish has only been taken to the fine diamond stage, but the resulting scratches do not materially affect the clarity with which the structure is revealed. Polarized light, unetched. × 100.

when rendered in colour by using the sensitive-tint plate, although colour does not always add any information of metallographic importance. Moreover, the identification of inclusions using polarized light may not be as positive and accurate as was once believed; the electron-probe analyser (see Chapter 11) is a much more positive (but very much more expensive) tool for this task.

The reader who has grasped the fundamental ideas outlined earlier in this chapter should have no difficulty in recognizing the potential uses of polarized light. There are two clear classes of effects—those arising from anisotropy and those which depend on oblique reflections. Each class may be subdivided. Under anisotropy the divisions relate to anisotropy of the metal (or alloy) itself, to that of non-metallic inclusions (especially

Fig. 87. Nodular cast iron, unetched. Structure of graphite modules revealed by polarized light. The dark crosses in the nodules arise in a manner similar to the isogyrs shown in Fig. 76. ×300.

transparent ones) or to that of thin layers formed preferentially on surfaces. Oblique reflections may be from features which have developed during deformation or phase changes and which are then studied for their own sakes, or from pits or grooves formed by some deliberate etching treatment.

Appendix

Anisotropic Metals; Etchants

The list below gives those metals which have been found to give grain contrast using polarized light. Section A contains anisotropic metals and Section B isotropic metals for which specific treatments have been developed to render them active to polarized light. These treatments are summarized in Section C.

A. **Anisotropic Metals** (crystal structure in brackets*)

Antimony	(Rh.)	Tellurium	(Hex.)
Beryllium	(H.C.P.)	Tin	(B.C.T.)
Bismuth	(Rh.)	Titanium	(H.C.P.)
Cadmium	(H.C.P.)	Uranium	(Orth.)
Cobalt†	(H.C.P.)	Zinc	(H.C.P.)
Magnesium†	(H.C.P.)	Zirconium	(H.C.P.)
Scandium	(H.C.P.)		

* Rh. = rhombohedral; Hex. = hexagonal; H.C.P. = hexagonal close-packed; B.C.T. = body-centred tetragonal; Orth. = orthorhombic.

† These metals may require etching to give good contrast (see Section C below).

Although they would probably require very careful preparation, it seems that several other metals including indium and thallium should be added to this list.

B. Isotropic Metals and Alloys

The number in brackets refers to the treatment described in Section C below.

Aluminium; Al-Mn, Al-Mg	(1)	Silicon-iron	(5)
Copper; also Cu-Zn, Cu-Sn	(2)	Pearlitic steels	(6), (8)
Copper-nickel (Monel)	(3)	Magnesium	(7)
Lead	(4)	Nickel	(8)

C. Etching Solutions for Isotropic Metals

1. Anodic oxidation is carried out in the following solution, on a previously polished specimen: orthophosphoric acid (85%) 53 ml, water 26 ml, "Carbitol" (diethylene glycol monoethyl ether) 26.5 ml, hydrofluoric acid (48%) 1 ml. The specimen is made the anode with 40V at room temperature for 1–2 min, stirring gently.[4]

   Alternative but similar formulae have been published by others.[5,6]
2. The polished specimen is lightly etched in alcoholic ferric chloride; time of etching 1–2 min; 2 g ferric chloride, 30 ml water, 10 ml hydrochloric acid, 60 ml ethyl alcohol.[7]
3. The polished specimen is immersed in a mixture of 3 g chromic acid, 10 ml nitric acid, 5 g ammonium chloride and 90 ml water. The time of etching was not specified.[8]
4. The polished specimen is lightly etched for 4–5 sec, washed, dried and re-etched two or three times in glacial acetic acid 7 parts, hydrogen peroxide (100 vol) 1 part, ethyl alcohol 1 part, water 2 parts.[9]
5. A solution of ferric sulphate of unknown strength has been used.[10]
6. 2 or 4% nitric acid in alcohol (nital).
7. Magnesium may be prepared to give enhanced contrast by etching for 6–8 sec, in a mixture of 100 ml (6%) picral, 10 ml water and 5 ml glacial acetic acid. The etching should be carried out on a freshly polished specimen.[11]
8. It has also been reported[12] that cathodic vacuum oxidation is capable of giving good contrast with polarized light for a wide range of metals. In addition to several already listed in Sections B and C are 0·08% carbon steel and 99·8% purity iron. The specimen is first etched with argon-ion bombardment for 3–5 min. Air is then admitted to a pressure of 5×10^{-2} torr and the oxide film built up using about 5 kV.
9. Anodic films on aluminium may be removed by etching for $1\text{–}1\frac{1}{2}$ min at 100°C in an aqueous solution of 35 ml orthophosphoric acid (85% strength) plus 20 g chromium trioxide per litre.[4]

REFERENCES

1 G. K. Conn and F. J. Bradshaw, *Polarized Light in Metallurgy*, London, 1952 (Butterworth).
2 E. C. W. Perryman and M. Lack, *Nature*, 1951, **167**, 479.
3 P. Lacombe in *Metallography—1963 (Proceedings of Sorby Centenary Meeting)*, London, 1964 (Iron & Steel Institute).
4 P. R. Sperry, *Trans. A.I.M.E.*, 1950, **188**, 103.
5 A. Hone and E. C. Pearson, *Metal Prog.*, 1948, **43**, 363.

6 A. Hone and E. C. Pearson, *ibid.*, 1950, **58,** 7113.
7 L. Barker, private communication.
8 D. H. Woodard, *Trans. A.I.M.E.*, 1949, **185,** 122.
9 R. C. Gifkins and J. M. Nicholls, *J. Inst. Metals*, 1959–60, **88,** 96.
10 P. Dunsmuir, *Metallurgia*, 1950, **41,** 240.
11 P. F. George, *Trans. A.S.M.*, 1947, **38,** 687.
12 F. Hilbert and H. Lorenz, *Jena Review*, 1965, **5,** 218.

8 Photography

The recording of microstructures by photography is the final stage in a series of processes involving skill and complex equipment; too often the result is made less effective as a means of conveying information because of poor photographic technique. There is no excuse for this. Modern photographic material is reliable and supplied with complete details for its proper processing; the only stage which is difficult is that of assessing the correct exposure. However, even here it is becoming less a matter of judgment and more one of applying the proper procedures and using appropriate equipment. Many metallographs have more-or-less elaborate accessories for determining exposure and even, possibly, for making the exposure automatically. It depends upon the principle used in the exposure meter whether this is a reliable method or not. The calibration of a microscope to give standardized exposures is quite simple, although a little tedious; it will be described later in this chapter.

The Aim of Photographs

The other point of general application in recording microstructures is that development of some skill in selecting and arranging the picture is a good investment. The aim of a photograph is primarily to communicate information and, in the majority of cases, it has to do this by some selective sampling. Moreover, it seems a universal truth that any method of communication degrades the information which is being conveyed: detail is lost, blurred or obscured. This, of course, is the idea familiar to scientists, that entropy increases unless positive steps are taken to organize the elements of a system. In the present example the positive steps concern clean technique and careful selection. An implied criticism is not, therefore, necessarily the point of the story of the metallographer who emerged from the microscope room tired and dishevelled, with the remark that he had spent the whole morning examining a specimen to find a typical area!

Photographic Materials

Some general remarks about photographic emulsions will be given before discussing those specifically suited to metallographic work.

Image Formation

A photographic emulsion (which is not truly an emulsion, but a dispersion) consists of small crystals or *grains* of a sensitive material, usually silver halides, held in a transparent medium (gelatin). A proportion of these grains becomes sensitized by the absorption of photons when light falls on the emulsion and thus a *latent image* is produced. This image can be *developed* by the action of suitable chemicals; these have the property of reacting with the sensitized grains more readily than the rest, converting them to small particles of metallic silver. If the development is prolonged, it eventually reduces all the halide grains to silver, producing a more-or-less uniform blackening or *fog*; even with normal developing times, there is some fogging. For this reason conditions are designed, and recommended, to minimize fogging to an acceptable background level. After development, the excess unreduced halides have to be removed, for subsequent exposure to light alone, without further chemical development, will slowly reduce the halides to silver and also produce fogging. This removal is effected by immersion in a fixer (a solution of "hypo"—sodium thiosulphate). Finally a thorough washing in water is required, to remove traces of the chemicals which might eventually oxidize and stain the emulsion.

What has just been said has carried the implication that it concerns only negative materials, in which the primary image is obtained with reversed contrast to that in the object photographed, but it applies equally to positive material (usually photographic papers) on which the negative is used to form the final record.

Photographic Density

The blackness of the photographic image is measured by determining the ratio of transmitted to incident light I_t/I_o; this ratio is the *transmittance* and its reciprocal is the *opacity*. The logarithm of the opacity is known as the (optical) *density*. This use of the word was brought in by two pioneers of photographic development, Hurter and Driffield, who also extended the concepts to the description of photographic emulsions in the *characteristic* or *H. and D. curve*.

Characteristic Curve—Gamma

The characteristic curve is obtained by plotting density D against the logarithm of total exposure E (i.e. the product of light intensity, I, and the time of exposure, t). A curve of the form shown in Fig. 88 is obtained. It may be divided into four parts: (1) the toe AB in which the general trend is one of underexposure and in which only a little contrast above background fog is obtained, even for the maximum change in exposure; (2) a straight portion BC in which the maximum changes in density for a given increase in exposure are recorded; (3) a shoulder CD, which is a region of overexposure and, as in (1), of low contrast; (4) beyond the shoulder a drop along DF, with an increase in exposure and then a decrease in density; this effect is known as *solarization*.

Two particular characteristics of the curve are of special use. The first is the slope of the straight portion (2), that is, the tangent of the angle α; tan α is called "gamma" and usually it is written as such, although the

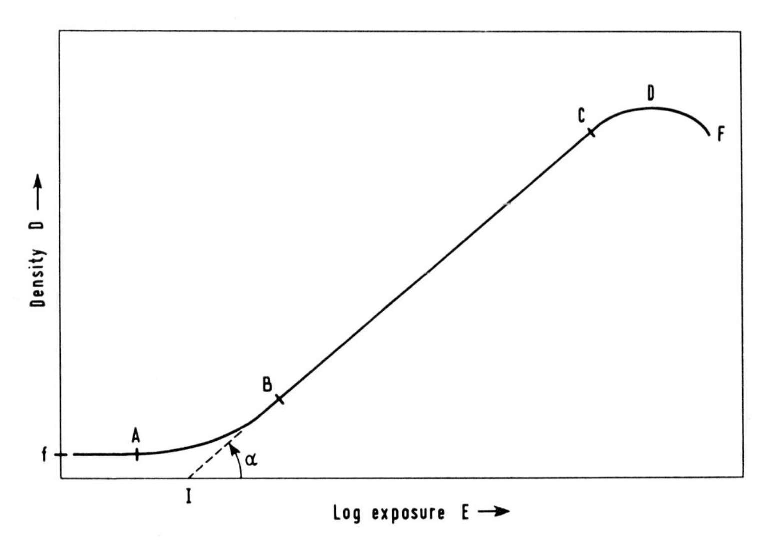

Fig. 88. Characteristic curve for a photographic emulsion in which (photographic) density D is plotted against log total exposure E, showing (1) a toe region of underexposure AB, (2) straight-line portion BC of correct exposure, (3) a shoulder region of overexposure CD and (4) a further overexposure along DF giving decreased density —*solarization*; the tangent of the angle α is called *gamma*; the value of log E at A is known as the threshold; the density f below the threshold is the *fog* level and I is the *inertia* point.

Greek symbol is also quite generally used. It can be seen that gamma describes the range of densities, or contrast, which can be obtained. The other parameter which is used is the value of log E at which the straight-line portion CB meets the exposure axis at I; this is usually referred to as the *inertia* or *sensitivity* of the photographic material. Clearly, it defines another threshold value of E where contrast above the background fog becomes perceptible and could be used to indicate the *speed* of the emulsion. Hurter and Driffield used this approach to characterizing speed many years ago (hence the term H. & D. Number), but it has been superseded by a series of methods which denote speed more realistically and which will be described presently.

It is important to realize that the characteristic curve depends not only upon the photographic emulsion, but also upon the conditions of processing. To mean anything useful the curve has to be linked to a particular developer and to a stated time and temperature of development.

Fig. 89 reproduces a typical set of curves, such as would be supplied by the manufacturer of the photographic material. It will be seen that the slope of the characteristic curve becomes greater as the time of development (at a particular temperature) is increased.

There is one other factor which may influence the interpretation of the characteristic curve, but which has not been mentioned explicitly. We

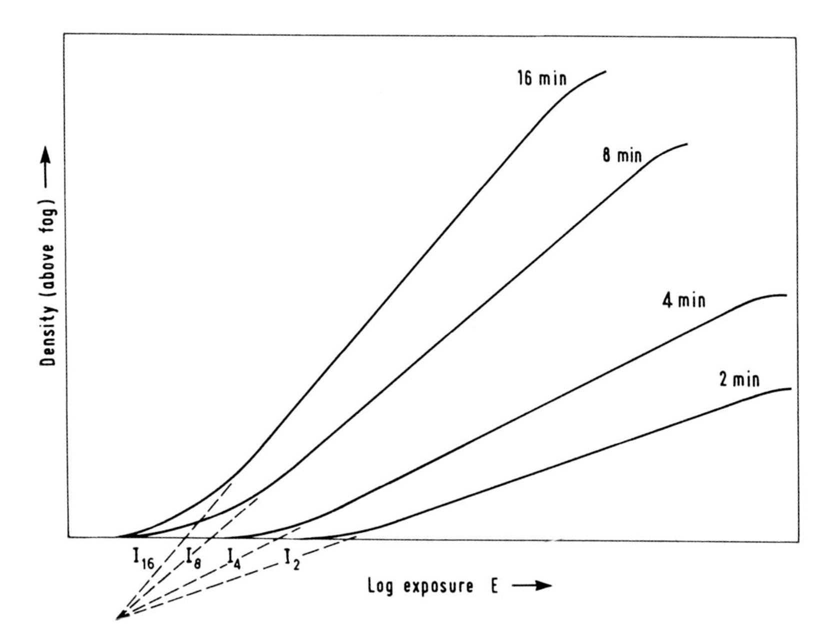

Fig. 89. Characteristic curves for a photographic emulsion for various development times (2–16 min) at a particular temperature and using a specified developer. Note that the straight lines meet below the E axis and I moves to the left as development time is increased. (Schematic.)

have been referring to the total exposure E, but defined this as the product of light intensity and time, It. The density does show marked variation from the normal curve if either I or t is very large and the other very small, that is, a very intense light acting for a very small time or a very weak light acting for a long time. This is spoken of as the *reciprocity law failure*, the reciprocity law itself stating that the amount of a photochemical reaction is dependent solely upon the actual light energy absorbed, not on its rate of absorption. The effect of the reciprocity law failure is only likely to be encountered in metallography when very long exposure times are used. The only case the author can think of was met using multiple-beam interferometry, when the conditions were not truly suitable for production of fringes; the reciprocity failure was not then a problem, because only black and white regions were being recorded (i.e. there were no intermediate tones).

Film Speed

The characteristic curve has been used in many ways to attempt to denote film speed. Of the five systems listed below, that of fractional gradient is the most widely used, but it is perhaps worth knowing how other systems are worked out.

1. Threshold—the speed is related to the value of E just sufficient to give a perceptible density above background fog. A major disadvantage of this method is that it bears no direct relation to the straight-line portion of the curve. It was widely used, however, in the Scheiner system.
2. Fixed threshold—the value of E required to give a fixed density (say 0·1) above fog. This was, for many years, the standard used in Germany by the DIN rating. It only partially overcame the shortcomings of the threshold systems.
3. Inertia—as mentioned earlier, the H. & D. rating was given by the reciprocal of the value of the inertia; the original Weston ratings (*now abandoned*) were essentially derived from this parameter.
4. Minimum useful gradient—this system attempts to compromise between the use of the toe and of the straight line alone, by taking a

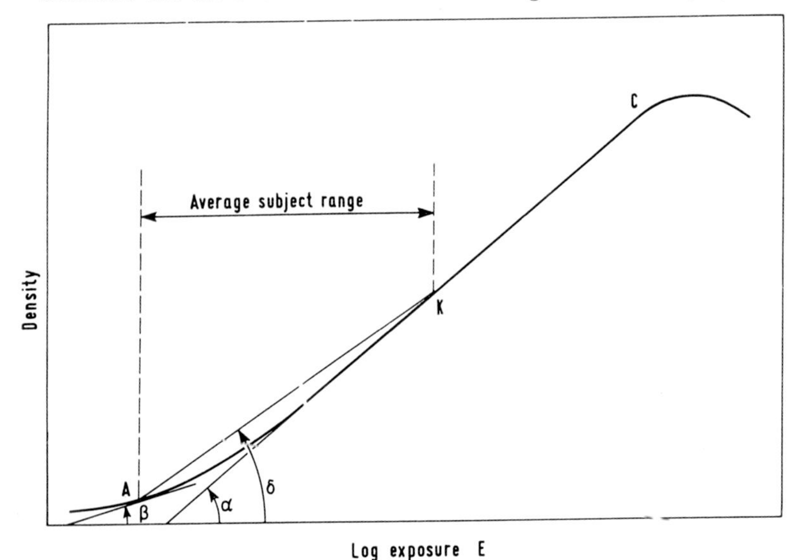

Fig. 90. The use of the characteristic curve in specifying film speed by the "fractional gradient" criterion (see text).

value for E such that tan α becomes 0·2—see Fig. 90. This method of rating was never used commercially, but led to the next method.
5. Fractional gradient—the concept here is to match a point on the toe to a given fraction of the gamma (slope of the straight line), chosen in such a way that it covers the contrast range of an average subject. Specifically, in Fig. 90, the slope of the tangent at A on the

toe equals a given fraction of the slope of AK, K being the point defined by the upper end of the contrast range. Thus $\tan \beta = \text{const.} \times \tan \delta$. In practice it has been found that if the constant is 0·3 and if $\tan \delta$ is obtained over a range of log exposures equal to 1·5 it is then possible to obtain "excellent" rather than merely "acceptable" prints from negatives exposed to meet the criterion. It is important to note that the whole concept of the fractional gradient method of denoting speed turns on being able to vary the minimum contrast available in the "shadows" and then building the contrast range upon this point; it is only possible to do this in practice because there are various grades of paper for printing the negative and these can be selected to give an excellent rendering of the shadow contrast.

The value of E at the point A is then taken to give the A.S.A. film speed as $1/E$. Dividing this by 4 gives the *exposure index* in the A.S.A. system for calibrating exposure meters. The exposure thus calculated gives somewhat greater than the minimum exposure to yield a negative capable of giving an excellent print, but it must be borne in mind that the indices have been computed with outdoor photography in mind. Too much reliance should not therefore be placed on them when both the type of "scene" and the lighting are as different from this standard as they are in metallography.

Tables giving working conversions between the various speed systems are available (see, e.g., *Ilford Manual*,[1] p. 289). From what has been said about their determination, it is clear that there can be no real correspondence between them, but it may be useful to have an approximate conversion.

Many exposure meters carry a scale in terms of *light values*; these are for the convenience of users of normal cameras, and give a parameter which combines aperture of the lens ("f" number) and exposure time. They are not, therefore, appropriate for use in metallography.

Determination of Gamma

It is worth determining the gamma of materials used in metallography under the conditions and with the processing methods for this particular work. To do this, it is necessary to have a calibrated step wedge, which is a glass plate having 10–12 steps ranging from clear to a density of ~ 3. The step-wedge plate is held firmly over the negative material with a piece of black paper as backing (to absorb scattered light) and then exposed with the normal illumination of the microscope to give a record of as many of the steps as possible after development. The development must be standardized.

The densities of the steps on the negative are then measured by determining the proportion of a given intensity of light which they transmit.*

* Photometric instruments for this purpose are available: similar instruments are used in spectroscopy. With a sufficiently intense light source and appropriate screening it is possible to use a normal photographic exposure meter as the measuring device.

It is thus possible to plot a characteristic curve of density of the negative against density of the standard (as in Fig. 88). From this the gamma is obtained for the conditions used. The process is repeated with a range of development times and so the variation of gamma with time of development plotted; the result would be similar to Fig. 91 (*a*). Should the

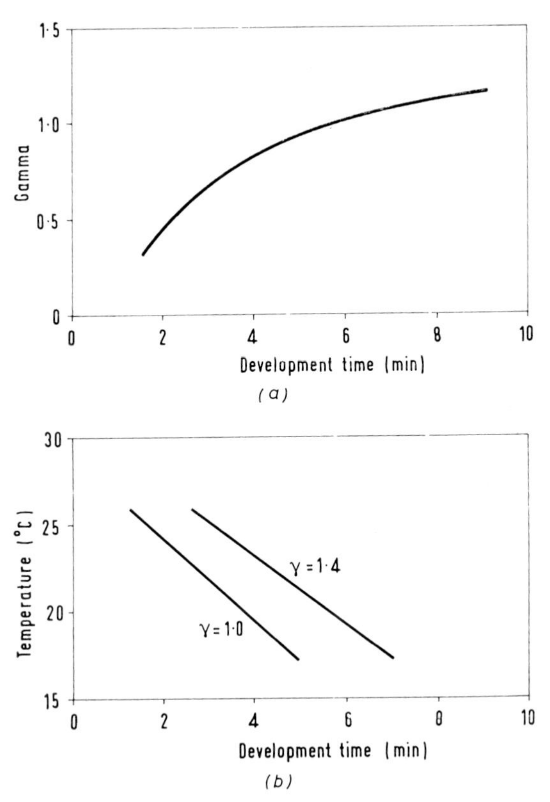

Fig. 91. Working diagram giving dependence of gamma upon time and temperature for a particular photographic emulsion and developer.
(*a*) Temperature fixed: film, RP99; developer, G2; temperature, 70°F.
(*b*) Temperature varied. (Schematic.)

temperature of development also be a possible variable (that is, if the developing tanks are not temperature-controlled) a series of lines showing the combinations of time and temperature to give various values of gamma may be determined (Fig. 91 (*b*)).

Plates, Films

It has been traditional to use plates (i.e. glass-backed negatives) in metallography, but these offer no advantages over film and suffer from disadvantages in their bulk for storage and potentiality for breakage.

Film may be in the form of separate sheets, known as cut film, or in the 35 mm rolls more familiar to the amateur photographer. There are advantages in using larger negatives than 35 mm frames; it is easier to focus sharply with the larger sizes and they can be developed more rapidly and yet not show "graininess" in the silver image. In a 35 mm negative this might be troublesome because it would obtrude when an enlarged print was made. And thirdly, of course, the larger size of negative can be printed directly by contact with printing paper without the use of an enlarger.

Graininess in 35 mm negatives can be avoided by the use of suitably formulated developers, but these may introduce other disadvantages because they tend to give a slower speed to the film and take longer to develop the image.

On the other hand, partly because it is a smaller size and partly because of the enormous bulk of material required in the cinematograph industry, 35 mm film is comparatively cheap. It is used widely, moreover, on microscopes which are not full-scale metallographs, but which are sufficiently robust to carry attachments consisting of the backs from normal 35 mm cameras.

Film or plates suitable for metallography are generally those capable of being developed to an acceptably fine grain size with a gamma of 1–1·5. It will be found that the gamma obtained with yellow-green light, as normally used in microscopy, will be a little higher than that quoted by the manufacturer, who will have used "mean noon daylight". Some firms make emulsions specifically for metallography. Any emulsion of reasonably high contrast is usually adequate; they will be described as "orthochromatic" or "process' and often have these terms combined with the adjective "commercial". A high-speed emulsion is not necessary and may even be a disadvantage because shutters on metallographs do not usually operate to give very short exposure times ($< \frac{1}{2}$ sec). Also, from what has been said about microscope objectives and from the use of yellow-green or blue filters, it will be clear that the colour sensitivity of the emulsions used need not be very wide. Photographic emulsions are modified by incorporating dyes into the gelatin so that the silver salts may be rendered sensitive to various wavelengths of light. In the undyed state the sensitivity is greatest in the blue and ultraviolet and very small in the red regions of the spectrum; plates with correction to the yellow region are called "orthochromatic" and to the red, "panchromatic". Although full colour correction is not required with photography of microstructures, it might sometimes be useful in macrophotographs; for example, in order to show up an oxidized area on a fracture surface. The section on colour photography takes up this point again.

Exposure Meters

The light intensity in the image obtained in a metallurgical microscope is not very high, so that the conventional photoelectric exposure meter is not very suitable unless some kind of amplification is used. Instruments to do this are available and, indeed, are built into some makes of microscopes. This kind of procedure measures the average intensity of light at the plane of the image; this might be appropriate to a specimen in which the structure is uniformly bright, but it will not necessarily give an exposure which allows a structure, in which there is a high contrast, to be recorded within the latitude of the emulsion used. It should be clear from what has been said about the characteristics of photographic emulsions that the aim of the exposure should be to get the extremes of contrast in the subject on to the straight-line portion of the characteristic curve.

Let us consider a specific example, that of a structure of ferrite and pearlite at a modest magnification, so that the pearlitic areas are dark and unresolved. An exposure obtained from a meter reading of the average intensity of light will give a negative that may be printed satisfactorily, by adjusting the grade of printing paper to give a slightly grey tone to the ferritic areas. However, the tone of the pearlitic areas will then be automatically fixed by the contrast available in the negative and thus will give a good representation of the original subject only if the extremes of contrast have been exposed on to the straight-line portion of the characteristic curve. The average exposure does not necessarily guide us to comply with this condition, although it might be possible to offset the errors by choosing appropriate printing conditions. This is often done. However, it is desirable, and quite feasible, to standardize exposures on a particular microscope for various groups of alloy types and thus also reduce printing to a standard set of conditions. It follows from what we have just said about the case of a specimen containing ferrite and pearlite that the way to standardization is through exposures based on the high-light intensity (i.e. upon the intensity from the ferrite) and for which it is ensured that the pearlite is recorded on the straight-line portion of the characteristic curve. Once this has been done, the exposure will be correct for all specimens containing ferrite and pearlite, in whatever proportions (provided that the etching conditions have also been standardized). It has been shown by Samuels *et al.*[2] that groupings of alloys can be made on a much wider basis than would be indicated by considerations of varying the proportions of particular constituents. Standard exposures can be derived for groups such as steels, aluminium, copper, bronzes, etc. Their method will be described in the next section.

Standardized Exposurers

It is recommended that the test exposures should be made by cutting a slot in the dark slide of a film holder, moving this along and so enabling successive strips of the film to be exposed for a series of durations. The simpler alternative method of withdrawing the dark slide completely and then replacing it step by step is not entirely satisfactory, because a number of incremental exposures does not give the same photographic effect as the equivalent single exposure.

There are two methods which can be used to process and assess the exposure, using an exposed test strip made with a polished unetched specimen of the metal concerned (say, mild steel for the steel group). The first is an accurate method employing a densitometer. The test negative is cut into three strips, each containing the range of exposures; one strip is developed to give an estimated gamma of 1·3, and the other two to give gammas a little higher and a little lower than the first. The density of each step is then measured in all strips (as in the construction of the characteristic curve described earlier in this chapter) and the results are plotted to give characteristic curves of density against the logarithm of exposure (Fig. 92). The three curves should intersect at the foot of the curves at A, so that the line representing an accurate gamma of 1·3 can be drawn through A—the broken line in Fig. 92. The maximum density

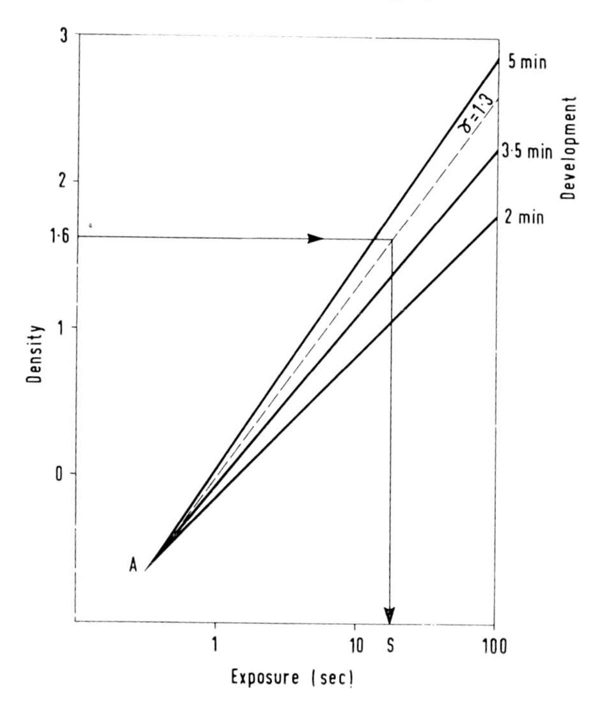

Fig. 92. Construction of a working curve for the determination of exposure for a particular type of metallurgical specimen. Density exposure curves for various times of development are used to construct the line for $\gamma = 1{\cdot}3$. A chosen high-light density (1·6) then yields a standard exposure *S*.
(After Samuels *et al.*[2].)

required in the negative, corresponding to the high-light (that is, ferrite in the case of a normalized mild steel) is chosen—see Fig. 92—and the standard exposures then read off from the curve for a gamma of 1·3. This density is chosen as low as practicable, in order to keep exposure times reasonably short. Samuels *et al.* found that the most contrasty ferrous

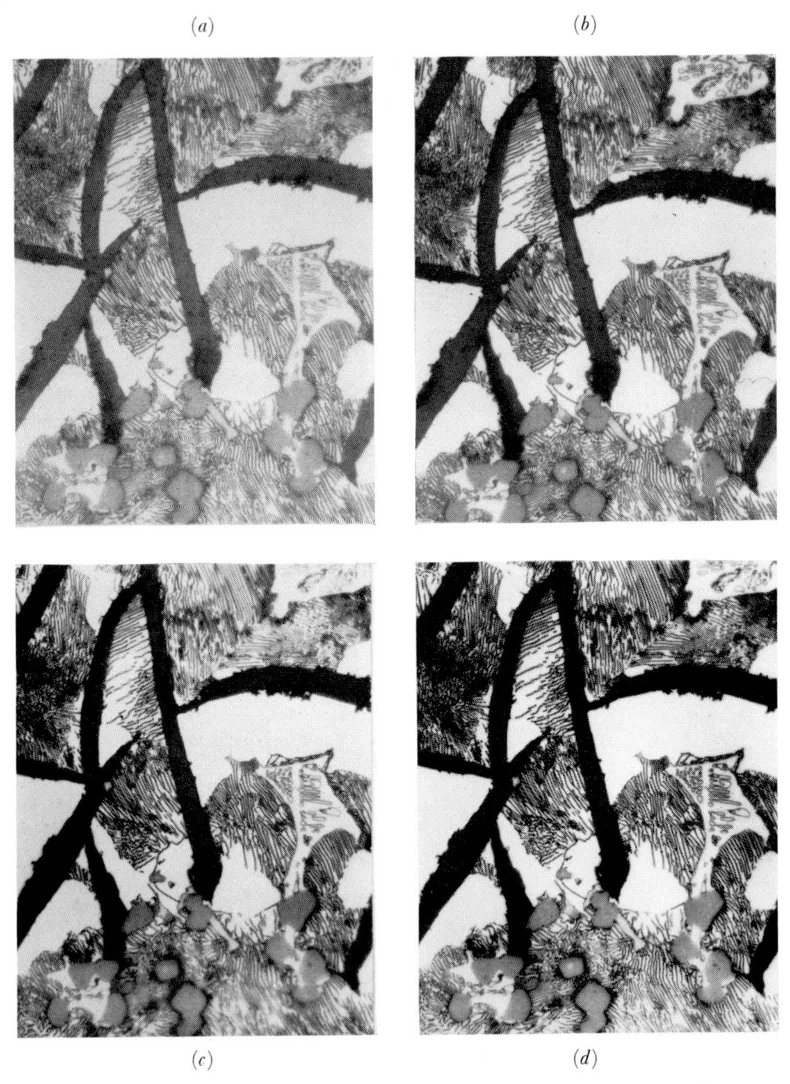

Fig. 93. Prints, on a normal grade of paper, of a series of negatives prepared to determine the correct exposure time for high-light density. The "best contrast" print—here print (*c*)—is selected to determine high-light density. This microstructure gives a full range of contrast, from the ferrite, resolved pearlite and flake graphite; etched. ×500.

Print	Exposure Time (sec)	Negative Density (ferrite)	Printing Time (sec)
a	5·2	0·9	0·8
b	8·2	1·1	1·4
c	20	1·6	3·8
d	48	2·1	14

specimens were those containing graphite or unresolved pearlite in a ferritic matrix, and these could be recorded satisfactorily if maximum density was taken as 1·6 *for the particular negative material used* (Australian Kodak "Commercial Orthochromatic").

In the second, less accurate, method a specimen having a high contrast is used as the subject for the step negative (or series of negatives). A normalized mild steel (etched) was found to be suitable. The negatives are developed identically to the recommended contrast for the emulsion used. From each negative or from the various test strips, prints are made on the normal grade of printing paper, the printing exposure being made to give the best contrast for each negative. The negative which then gives the best contrast (without unreasonably long exposure time in printing) is selected and its high-light density taken as the standard. A series of such prints is shown in Fig. 93 (*a–d*). On the basis of what has just been said, print (*c*) would be selected and the negative from which it has been made taken to determine the high-light density.

It was found by Samuels *et al.* that metals and alloys could be put into groups such that each group would have a particular standard exposure. They determined the relative exposures for these groups, so that all that is necessary now is to determine the standard exposure for one specimen, as described above, and then use their conversion factors shown here in Table 2.

TABLE 2

Exposure Factors for Alloy Groups

Metal or Alloy	Exposure Factor (Al = 1·0)
Al; Ag; Zn; Cd brass (10–45% Zn) .	1·0
Sn; Sb	1·2
Cu; Ni; brass (0–10% Zn) . . .	1·3
Sn or Al bronze; steel . . .	1·5

Other exposure factors may be required. If the aperture stop is not fully opened, this will demand increased exposure to produce a standard negative. Samuels *et al.* found that the exposure factor e was given by the equation:

$$e = 4 - 0{\cdot}03\, A$$

where A is the percentage of the aperture diameter opened.

It might also be necessary to introduce factors to allow for variations in the intensity of light from the lamp, or from lamp to lamp. However, it was found that carbon-arc lamps remained very constant over a wide range of currents (7–10 amps) and that tungsten filament lamps of a particular make gave reproducible intensities provided the operating voltage did not vary by more than $\pm$ 0·1 V from the recommended voltage. Fortunately, the major deterioration of a tungsten lamp occurs in the first hour or two of use. Xenon arc lamps are becoming increasingly used on metallographs, and these add to their other advantages that of excellent stability.

Exposure factors may also be required for other changes—for example, different kinds of oblique illumination, phase contrast and so on. All of these should be found as they are needed and incorporated into a table which shows the basic conditions and corresponding exposure factors as in the sample set of results in Table 3. A subsidiary table should be compiled to show factors associated with changes in types of illumination, etc.

TABLE 3

Calibration and Exposure Table
Research Microscope No. 3

Magnification	Objective	Ocular	Aper. Stop	Bellows Extn.	Exposure Factor for Various Groups			
					1·0	1·2	1·3	1·5
100	12·5X Achro (N.A. 0·20)	5X Hyg	3·0	210	4·5	5·5	6·0	7·0

A point which has been adequately demonstrated in using the above approach to standardization is that apparent failures in the system are almost certainly due to some breakdown in the standard conditions used; that is, if it is not the illuminating source which is at fault, it is almost certainly something in the dark room, such as worn-out developer or lack of temperature control of developer. There are occasions when the specimen is outside the contrast range used in the calibration, but these should be recognized easily with a little experience.

Printing Papers

We have already jumped ahead and spoken of grades of printing paper in the previous section; we should now, perhaps, look at this stage of the photographic process more closely. Emulsions basically similar to those used for negative materials are employed in the production of positive or printing materials; they are usually on a paper backing. There are two principal types, contact and enlarging papers; papers may also be classified according to the kind of photosensitive material used. Bromide and chloro-bromide papers are the faster and so more convenient to use when the light intensity is lower, which is often the case in enlarging. The slower chloride papers are used for contact printing. There is no reason why bromide paper should not be used for contact printing; many photographers prefer to do so.

Chloride papers are the slowest and have the finest grain; they are used for contact printing (hence they are generally called "contact papers"), the negative being held with its emulsion side against the emulsion side of the paper and exposed to light. The colour of the image of a chloride paper which has been fully developed is black, but some manufacturers incorporate a blue dye in the emulsion, yielding a less "warm" colour, namely that of blue-black. Since the speed of chloride papers is relatively low, they can be handled in a reasonably bright yellow or orange light from a safelight, as specified by the maker. Development times for contact papers are about one minute at 68° F (20° C).

Bromide papers (or enlarging papers) form the other main class; they are about fifty times as fast as contact papers and coarser grained. However, the graininess of paper emulsions is rarely, if ever, obtrusive. Any graininess in the final print will almost certainly arise from the negative, but this is also unlikely in metallographic work, where high-speed films are not demanded. As the alternative name implies, bromide papers are used for making enlargements; their greater speed enables prints to be made with reasonable exposure time using the relatively low light intensity available from an enlarger. Because of their high speed, the dark-room light they will tolerate without fogging has to be less than for chloride papers and either orange or greenish-yellow in colour. Development time for bromide papers is usually about 2 minutes at 68° F (20° C). They then give an image of a neutral black colour, but can be given further treatments ("toning") to convert them to blue-black, sepia, blue, yellow and other colours.

Chlorobromide papers form a third class; they have emulsions containing both chlorides and bromides of silver, the proportion varying from type to type of paper. They tend to have rather greater latitude in development than the other classes of paper, but the tone of the image depends upon development so that neutral black images are not generally obtained.

PAPER GRADES

Papers are graded according to their contrast range, having grade numbers 0–6; "0" is the "softest", that is, has the smallest range of contrast, grade "6" is the "hardest" or most contrasty. The contrast is controlled, of course, by designing the emulsions to have various specific values of gamma together with appropriate lengths of the straight-line portion of their characteristic curves.

Let us consider what it is we are trying to achieve in matching the grade of paper to a particular negative. Suppose the contrast range of the negative is such that the ratio of the highest and lowest opacities is 25:1. To print this without loss of tone values requires a paper which has gradation in the same ratio. By this, it is meant that the paper will just become darkened by exposure to a given light source after (say) one second and will give maximum density or blackness after 25 seconds. If the negative is printed on this paper with 25 seconds exposure (or perhaps a little less in actual practice), the paper then receives sufficient exposure to produce a dense black tone from the lightest parts, but the densest parts of the negative do not transmit enough light to cause any tinting of the paper at all.

The effect of printing a negative with a much smaller range of densities, say 15:1, would be to give a print with tones ranging from white to some tone of grey instead of the full range from white to black. An attempt to cure this by increasing the exposure time would, of course, result in giving a range of tones from grey to black.

The availability of grades of paper allows negatives of widely varying contrast ranges to be printed to the best advantage, giving adequate contrast in the final image, but the aim of standardized exposures is to obtain such a print with a reasonable exposure and full development using a "normal" grade of paper—i.e. grade 1 paper.

However, the more contrasty grades of paper are indeed available and should be used if they assist in making detail clear in the final print, without actually suppressing any features by clogging the denser parts or moving them into the general area of high-light.

From what has been said above, matching paper to negative requires a knowledge of the negative's range of densities. If the means of measuring this on the negative is readily available, it should be done, at least until skill in judging the range has been acquired. The range is characterized by one reading for the high-light areas, say 1·4, and one for the shadow, say 0·2; the density range or scale is then 1·2. The best paper for printing the negative is taken to be the grade having a density scale of 0·2 greater than that for the negative, in this case $1{\cdot}2 + 0{\cdot}2 = 1{\cdot}4$. Table 4 shows the grades of paper and their corresponding density scales, together with the density scales of negatives which would be best printed on each grade.

TABLE 4
Density Ranges of Grades of Paper

Grade	Density Range	Negative Range
0	1·7	1·4
1	1·5	1·2–1·4
2	1·3	1·0–1·2
3	1·1	0·8–1.0
4	0·9	0·6–0·8
5	0·7	0·6

PAPER STYLES

A glossy paper is almost always used for making metallographic prints; such paper has a smooth finish which is enhanced by glazing the emulsion after washing of the print is complete. The grain of glossy paper is unobtrusive and so the detail of the print is displayed clearly. Printing papers are made with a great variety of non-glossy finishes and also with tinted papers, but these are intended for artistic presentation of photographs rather than for scientific work.

Contact papers are invariably "single weight", whereas enlarging papers may be either single or double weight. Double weight is, of course, a thicker grade of paper, about equivalent to the thickness of a postcard. Lightweight or airmail papers can also be obtained.

Colour Metallography

There are many cases when etched structures, interference fringes and so forth appear to demand the removal of the green filter and use of colour film. The word "appear" is used here advisedly, for the lure of colour is strong, although it may not always convey any useful scientific information which cannot be recorded in black and white. However, there are cases when colour does perhaps add information; it may distinguish phases, identify fringes or reveal the whereabouts of a patch of corrosion product. Colour film is now reliable and, although more expensive than black and white film, not excessively so. Colour prints, on the other hand, are expensive, and their reproduction in journals and books is usually prohibitively so, but sometimes a combination of scientific point and aesthetic appeal may warrant the expense of colour reproduction.

Use of colour film is quite straightforward, and generally any intense source of light of reasonably wide spectral range is adequate; in this category would be included carbon arcs, special tungsten filament lamps and xenon arcs. It is often stated that precautions concerning the "colour temperature" of illumination are necessary in metallography. This is only relevant if, for some reason, the precise colour of a feature has to be recorded. Generally it is colour contrast that is required, and there is no point in pursuing perfection in the spectral rendition. The colour film will, in fact, record whatever colours the microscopist observes on the focusing screen and if these serve to record the scientific message this is all that matters.

The colour film normally used in snapshot cameras is reversal film, intended to be processed to give transparencies which are positives; to make prints then requires the preparation of a negative from the transparency. For metallographic work it is therefore preferable in most cases to use a negative colour film, from which the prints can be made directly. Such film is obtainable in cut sheets of sizes to fit normal film holders on metallographs. It has a further advantage that both processing the film and making the prints is less complex than for most reversal films, so that the metallographer may process his own material if he wishes. The advantages of colour transparencies for illustrating lectures should not be overlooked—they may often add a touch of artistic verisimilitude to an otherwise bald and unconvincing narrative.

It is also possible to record coloured characteristics in black and white by the use of selected filters which can emphasize the colours of interest, or bring out detail otherwise lost. A particular case which provides a good example is that of a diffusion couple of Nb–Sn, discussed by Love and Picklesimer.[3] They found that the variation in content of niobium could be shown up by an anodizing treatment, but the detail was lost in white light; suitable etchants to give a clear differentiation were not available. By carefully selecting two filters, they were able to take photographs with each of these in turn and so bring out in strong contrast the detail of the niobium concentrations in a pair of prints.

Quite often the colours to be photographed can be clearly differentiated on the final print by using an appropriate negative material. This does not necessarily have to be a panchromatic one; indeed, a more restricted range of chromatism might serve to enhance the contrast rather better than a panchromatic emulsion does. The metallographer should therefore select the material with a spectral response suited to the subject.

Before going to a great deal of trouble and expense in seeking accurate response to colour the merits of arrows stencilled or transferred on to the prints to indicate phases or features of interest should be considered.

Appendix

Polaroid Photography

Land, the inventor of the artificial polarizing sheet Polaroid, has also perfected a system of print-making which is very rapid; it is justifiably referred to as instant photography, the positive being ready some 20 seconds after the exposure of the negative has been made. The system uses processes of development, fixing and making of the positive which are specific to it. One principal difference from normal photography is that the positive is made by the diffusion of silver from the negative to the positive emulsion, where it is deposited in a colloidal form. The film is supplied, in roll or sheet form, with negative and positive emulsions capable of being brought facing each other but slightly separated. After the negative has been exposed, rolling on the film or withdrawing a flap for sheet film brings the two emulsions together with a layer of viscous developer between them; the developer is held in a sac which is broken as the juxtaposition of the emulsions is effected. It will be noted that another difference from normal photographic practice is that the positive emulsion does not have to be sensitized by light.

Special camera backs are made to hold the film and operate the mechanics of the development and printing processes. After the positive has been made, negative and positive are pulled out and detached; the positive is on a paper backing, and the emulsion may be protected by a glazing operation, whereby varnish is spread over it from a small supply held in a sponge.

The colloidal nature of the silver in the positive image renders the dark parts of the image intensely black and reduces graininess considerably. Many types of negative emulsion are available, including ones with long density ranges, panchromatic sensitivity or high speeds. The fastest emulsion available is very fast indeed, having an A.S.A. rating of 10,000; this compares with the usual speeds of 100-200 A.S.A.

If duplicate prints are required a special type of Polaroid film is required; the lack of this facility with the general types of Polaroid film is, perhaps, its major disadvantage. Colour photography is also possible using Polaroid film, which produces a colour print as rapidly as it does the black-and-white variety.

A recent critical examination[4] of the Polaroid process in metallography has concluded that there are difficulties in obtaining adequate contrast in prints made with achromat objectives. Since, as we have emphasized earlier, it is often contrast rather than resolution which it is difficult to achieve in metallography, this may be a serious shortcoming of the process.

The advantages of this system, which is competitive in cost for single prints, are readily imagined for many applications. Having a print available immediately permits corrections to exposure time, focus, contrast conditions and so forth to be made before specimens tarnish or without losing a selected field of the specimen; rushed reports may also be illustrated speedily. As so often happens with special methods in metallography, it is not recommended to replace more usual techniques, but to supplement them.

REFERENCES

1 A. Horder (ed.), *Ilford Manual of Photography* (5th edn), London, 1958 (Ilford).
2 L. E. Samuels, T. O. Mulhearn and R. M. Robb, *Metallurgia*, 1958, **57**, 342.
3 G. R. Love and M. L. Picklesimer, *Trans. A.I.M.E.*, 1966, **236**, 1505.
4 C. P. Johnson, *Metals Australia*, 1969, **1** (10), 333.

9 Miscellany of Aids to Metallography

1. Etching for Optical Examination

Desirably, a book on optical metallography should contain a discussion of the principles of etching specimens as well as recommended formulae. It must be admitted that no really satisfactory theory of etching exists, although it is clear that certain basic principles of corrosion theory are applicable. However, these are more useful in explaining why a particular reagent works rather than leading to the formulation of a reagent to perform a particular task on a particular alloy. In view of this, all that will be done here will be to list a few general points on etching and refer the reader to several compilations of etchants for various metals and alloys. In other words, there are two ways of finding out how to etch a particular alloy. The first is to look it up in one of the formularies.[1-3] The second is to develop a new mixture, starting with one that gives satisfactory results on some similar alloy. To assist in the search for new etchants, some attempt will be made to classify chemical etchants and list alternative ways of revealing structure.

EXAMINATION BEFORE ETCHING

There are many features which are best examined on unetched specimens. Indeed, it is good practice always to look at a specimen in the unetched condition in the first place, particularly when the specimen has been taken from a component which has failed in service or when it is being used to sample the quality of material. Non-metallic inclusions, cracks and cavities are obvious cases of features which should be viewed before etching; these remarks apply more particularly to mechanically polished specimens. The chemical attack inherent in electropolishing may have already done as much damage to the outline of cracks or inclusions as etching is likely to do, although it may be minimized by using the correct conditions. It must be realized that even in mechanical polishing there is an element—according to Samuels,[4] an overriding one—of chemical attack in final polishing when using media other than diamond pastes; thus, even with mechanical polishing, some rounding of cavities or cracks may occur unless the polishing is of the highest class. Whether this is significant must be judged in the context of the particular investigation.

DEPTH OF ETCHING

Generally the attack by etching should be kept to a minimum. The conditions of examination dictate what this level of attack should be. For macro-examination, etching usually has to be fairly deep in order to bring the features of interest into sufficient relief. As the magnification and resolution are increased, the depth of attack has to be reduced, for otherwise the coarseness of the attack could impose a structure (such as resolvable pits) upon the true structure of the alloy. Furthermore, a deeply etched specimen examined at high magnifications could also have changes of level outside the depth of field of the objective. With phase-contrast illumination the desirable depth of etching is usually even smaller, for we now rely upon the phase-contrast effect to obtain contrast and this reaches a maximum for steps of a few hundred Ångstroms.

The preceding remarks may appear trivial and obvious, but they do emphasize the fact that the optimum results are obtained only when there is some understanding of the physical processes involved in making observations.

TYPES OF ETCHING

(*a*) *Chemical Dissolution:* Probably the majority of etched specimens are prepared by dissolution of the specimen, the solutions used being selected because they dissolve various constituents differentially. Strong oxidizing agents form the principal group of reagents—these may be strong acids like nitric, sulphuric, or chromic, but usually they are diluted considerably. A group of organic acids, such as citric, picric, acetic and oxalic acids is also extensively used. These are comparatively weak acids and are used in fairly high concentrations. Quite often non-aqueous diluents are used, such as ethyl alcohol, glycerol or one of the more complex glycols. Their functions seem to be to dilute the active agent without allowing complex aqua ions to form with ions of metal or metal oxides. They have the effect, therefore, of moderating etching reactions which would otherwise lead to pitting and grooving, or sometimes merely to staining.

Other reagents do, in fact, give differential staining of constituents, presumably by depositing coloured or finely divided particles of some reaction product on the phases examined. Ferric chloride, hydrochloric acid (sometimes) and ammonium persulphate are examples of this class.

One would like to be able to make some generalizations about etching, such as when to immerse gently, when to agitate and when to swab the specimen, but it must be admitted that the subject does not seem to have become systematized sufficiently to do this.

(*b*) *Tinting:* There are several ways of obtaining tinted layers on specimens such that there is a differentation in colour (or depth of a particular colour) from one constituent to another. Controlled chemical oxidation or sulphide formation are the principal ones, and have been particularly developed by workers in France. Lacombe has given an excellent review of these methods and, indeed, of the whole range of etching techniques[5]. They have been shown to be capable of sensitively differentiating between phases. Fig. 94 shows a sulphided specimen of cast aluminium bronze.

Another kind of oxidation is obtained by heat tinting; as its name implies, the specimen is allowed to oxidize in air (or in a gaseous mixture containing a specified portion of oxygen) and oxides giving interference colours allowed to develop.

(*c*) *Electro-etching:* This is not essentially different from chemical etching. Reagents are chosen which dissolve phases of the alloy examined on the application of a current. As in the chemical case, metal ions are formed at

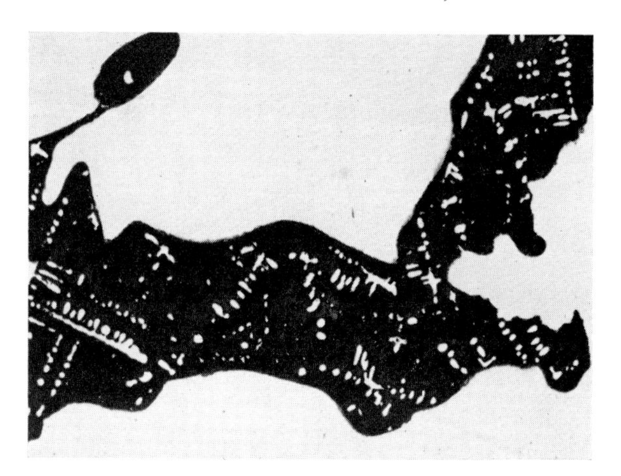

Fig. 94. An example of etching by sulphide tinting; cast aluminium bronze, showing segregated copper (dark) and copper oxide dendrites. ×50.

the anodes and electrons appear at the cathode. If, of course, some regions are more anodic than others (by virtue of their composition, segregation or local strain) they dissolve preferentially. The advantage of electro-etching is that it is capable of more precise control than the average chemical etchant.

The potentiostat is a device that carries this a step further. It enables voltage in the electrolytic cell to be maintained at a predetermined level, so that particular anodic (or cathodic) reactions may be isolated. In this way it is possible to etch various phases selectively. Once these voltages are known for a particular system, the voltage can be used to identify constituents.

A further variation on the basic theme of electro-etching is incorporated in commercially available electropolishing devices. It has the means of holding the specimen in a sealed glass cell through which the electrolyte is pumped. The specimen may be observed through the cell with a microscope whilst it is being etched electrolytically, so that the depth of etch or the varied response of the different constituents may be watched directly.

(*d*) *Vacuum Etching:* Grain boundaries, grains, twins and other features may be revealed by heating suitable specimens *in vacuo*. Structure is revealed by ridges or pits which are related to specific crystal planes. Fig. 95 shows the effect of vacuum etching silver.

(*e*) *Cathodic Etching:* This method is very useful for etching refractory alloys or specimens comprising layers of very dissimilar alloys, which would etch chemically at different rates. The specimen is made the cathode in a discharge tube and bombarded with ions of the gas in the tube. The simplest arrangement is to have the discharge tube in the form of a bell jar.

Fig. 95. Vacuum etching, used to show twins in silver and characteristic steps. ×750. (E. D. Hondros.)
(Reduced to $\frac{1}{2}$ in reproduction)

The specimen must be cooled efficiently, and therefore it has to be held in good contact with a water-cooled cathode. The anode is placed a few inches away, often above the specimen. The rate of etching is controlled by the energy and size of the bombarding ions and by the current density they impose, but it appears that more reproducible results are obtained with lower ion energies. A typical current density is 20 mA/cm^2, and a typical ion energy 5 keV. Many ions have been used successfully, the principal ones being those of the inert gases; obviously reactive ions would introduce chemical effects which would be likely to confuse the structural evidence sought. The bell jar is pumped out to 10^{-3} torr and the gas (say, argon) is admitted to a pressure of 10^{-2} torr. Times of etching vary according to the precise conditions, but times between 30 and 60 sec are reasonably typical.

As mentioned, the advantages of using cathodic etching are greatest with very refractory alloys or with sandwich specimens, but it has been used successfully for a very wide range of alloys. These include lead and copper alloys, aluminium bronzes, steels, titanium and chromium alloys. It has even been found possible to etch alumina, by placing a fine steel gauze over the specimen to provide the conducting cathode.

Fig. 96 shows an example of a cathodically etched specimen. Further information may be found in references 5 and 6.

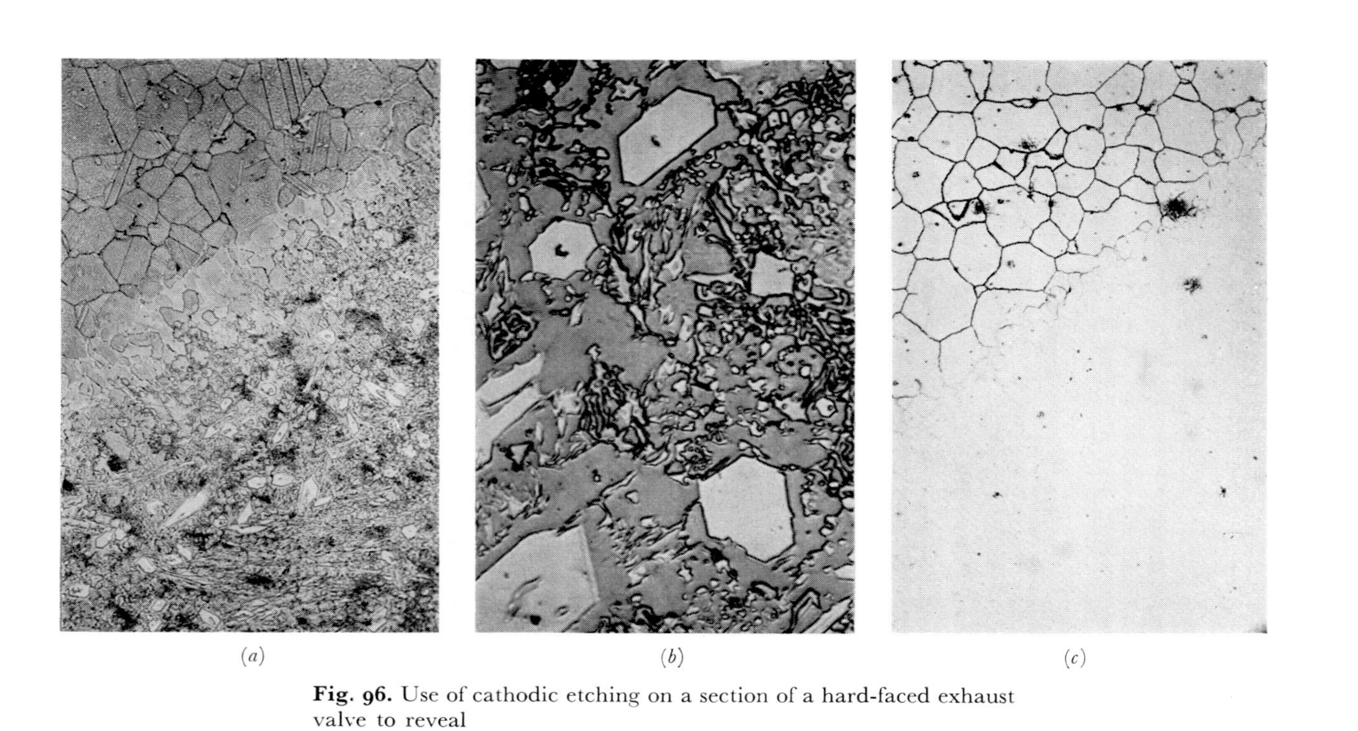

(a) (b) (c)

Fig. 96. Use of cathodic etching on a section of a hard-faced exhaust valve to reveal

(*a*) layers of dissimilar alloys, × 100;

(*b*) detail in the hard facing, × 1000.

Compare (*c*), in which an area is etched conventionally using hydrochloric acid, × 100.

(Reduced to $\frac{9}{10}$ in reproduction.) (After D. M. McCutcheon.)

(*f*) *Etch Pitting:* A special kind of chemically-etched surface may be obtained by using reagents that are very selective and only attack the sites of emergent dislocations. Crystallographic pits may then be formed and these give information about dislocation densities, sub-grain boundaries, orientation, etc. There is a vast literature on this subject[7] and the reagents do not always behave as described when used on slightly different (or even nominally identical) specimens. Fig. 97 shows typical etch pits in

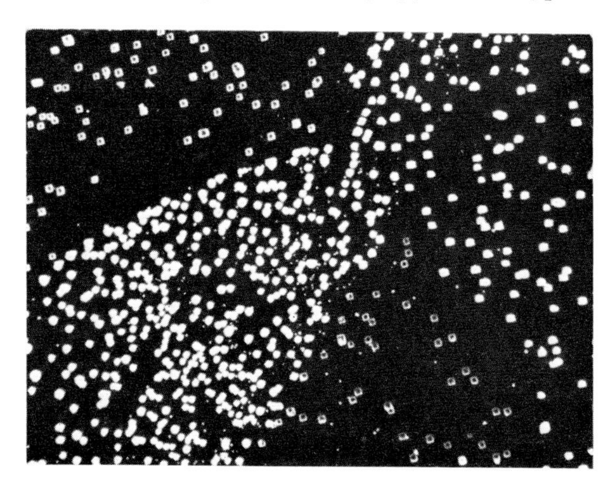

Fig. 97. Etch pits in aluminium, indicating grain orientations; dark-ground illumination, × 100.

aluminium, here revealed by dark-ground illumination. This technique shows the pits more clearly than bright-field illumination and aids when counting them; the reason for this is connected with visual acuity—see page 18.

2. Macrophotography

A great many macrostructures require so little magnification for their proper recording that the inclusion of a chapter on the subject in a book principally devoted to the microscope perhaps requires some further justification. Clearly, a good macrophotograph may be as important in telling a scientific or technological story as a microphotograph. Moreover, many macrostructures are indeed best revealed by a small magnification, often using special lenses which are fitted to the standard metallographs. On the basis of the primary magnification used, we may divide macrostructures into three classes:

(i) those reduced below natural size;
(ii) those with magnifications up to approximately × 5;
(iii) those with larger magnifications.

For classes (i) and (ii) a good camera with a focusing screen is required, together with some suitable lighting equipment. The problem almost always reduces to that of obtaining sufficient contrast combined with even illumination. Standard photographic floodlights, spotlights and fluorescent tubes mounted on stands, all have their uses and are, in general, deployed in the manner they would be in normal studio photography. One of the numerous handbooks on this subject would provide a good starting point for mastering the art of macrophotography, for it would describe the way shadows can be softened, the exposure calculated and so forth.

The most frequent cause of trouble is the appearance of high-lights or reflections on the specimen when it is otherwise satisfactorily lit to show good contrast. Photographing the specimen under water or kerosene, the

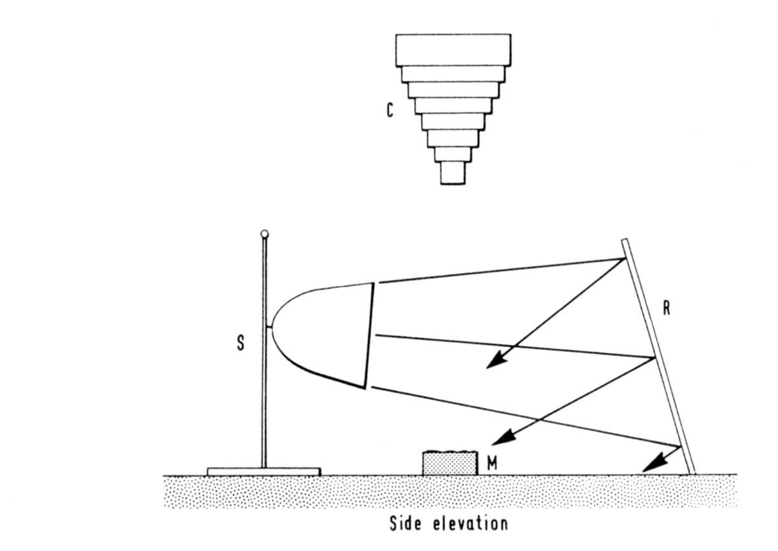

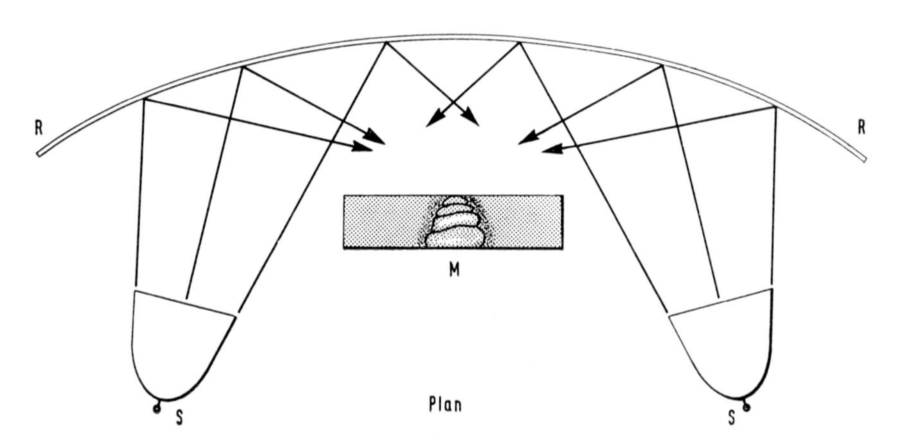

Fig. 98. Arrangement for obtaining even illumination for macrophotographs at low magnifications. Light from the lamps S S is scattered from the matte surface of a curved reflector R R on to the specimen M.

use of a polarizing filter, waxing or spraying the specimen surface* to make it matte, are all tricks that may be tried.

Very diffuse but oblique lighting is also often successful. This may be obtained by placing diffusing screens over the lights. Another way which often seems to give superior results is to place a large sheet of matte white card behind the specimen (Fig. 98) and illuminate the card fairly horizontally, so that only scattered light from the card reaches the specimen.

The foregoing remarks apply more particularly to macro-etched specimens which are otherwise flat. With fracture surfaces, manufacturing defects, etc., it is often necessary to obtain the principal moulding by an intense beam of light from one direction and then soften the shadows and give the specimen solidity by less intense or less focused beams from other directions. Fig. 99 shows an example of this kind of technique.

Fig. 99. Fracture surfaces of a bolt, taken using direct lighting, with the shadows softened by a second weaker source of illumination; a good three-dimensional effect is obtained. (J. Kelso.)

For class (iii), as has been mentioned, most large metallographs may be employed, using special long-focus objectives and also, sometimes, using modifications to the lighting system to obtain a broad oblique beam. The difficulties encountered are essentially those mentioned in discussing classes (i) and (ii) and the principles in overcoming them are also similar, although there is probably less flexibility in doing so. Fig. 100 shows an example of a macrophotograph taken in this way.

There are some instruments specifically designed for the examination of profiles of such specimens as gear wheels and other machine parts. These may have a range of magnification from $\times$ 10 to $\times$ 50 and suitable lighting

* Pressure packs of hair-setting spray have been found to be particularly effective.

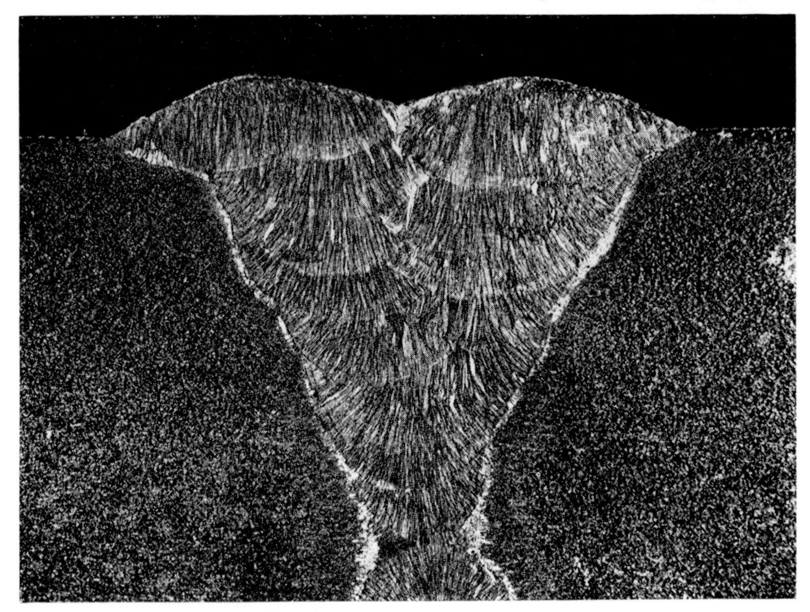

Fig. 100. Macrophotograph of a weld, taken using oblique illumination. $\times 4$. (W. Gifkins.)
(Reduced to $\frac{3}{4}$ in reproduction)

Fig. 101. Macrophotograph of a weld in copper taken using normal incidence lighting in a "profile projector". $\times 10$.

to obtain first-rate macrophotographs—see Fig. 101. It should be noted that these instruments are relatively expensive, so that it would not usually be economically sound to buy them solely for use in macrophotography, particularly since their lowest magnification is $\times$ 10; a great many macrophotographs seem to demand a magnification between $\times$ 10 and normal size.

Colour photographs are excellent in assisting the two-dimensional realization of solid, three-dimensional objects, although colour has a distinct economic disadvantage when reproductions are required for reports or papers. However, whilst this disadvantage is nearly always overriding for papers or even for books, the cost of colour prints, made from colour negatives and not from positives, is such that a limited number of copies for internal reports or reports to customers would not generally be excessively expensive. The kind of feature best rendered in colour often arises from the operation of corrosive conditions, as in some kinds of fatigue. A fracture surface may have coloured zones which show the progress of rupture and emphasize the role of fatigue. Studies of oxidation may also give specimens which demand colour photography for their most convincing reproduction.

3. Replicas for Metallography

There are occasions when it is worth making a replica of the surface of a micro-specimen. Faithful and stable replicas may now be made very easily, the techniques for this having been developed primarily for electron microscopy. The most obvious cases for this treatment are those where the area of interest is somehow inaccessible to the microscope. This may be because of the limited working distance of objectives, particularly those of higher N.A., or it may be that the object to be viewed is too large to put on a microscope stage. There are also occasions when a micro-examination of some material in the field is required; if a suitable surface can be prepared *in situ*, a replica may then be easily taken from the field to the laboratory. The other class of use for replicas is concerned with making a record of a specimen which is undergoing changes, e.g. during creep or fatigue. Replicas may be made at various stages of a test and thus detail of the entire specimen preserved. In this way, the earlier history of any areas which turn out to be of particular interest may be traced.

A great variety of plastics and techniques of using them have been developed for use in electron microscopy; these are very fully documented in a book by Kay.[8] For optical metallography, where the application is perhaps less critical, the simplest techniques suffice.

The simplest method of obtaining a replica is that using cellulose acetate in sheet form, usually of the order of a half-millimetre thick. A piece of the required size to replicate the specimen or area of interest is cut out and washed thoroughly in a degreasing solvent such as ether; it is then immersed in a shallow dish of ethyl acetate. When the film has softened it is picked up with tweezers and dropped on to the specimen; it is then immediately pressed firmly into contact with the specimen, either using the thumb or some form of squeegee. After some 5–10 minutes the film should have dried sufficiently for it to lift off when a corner is prised free. Gently drying under a reading lamp may accelerate the process. The principal causes of poor replicas are dirt or scale on the specimen or air bubbles trapped under the film.

The dry replica, when viewed with bright-field illumination, will give reasonable contrast. This may be improved by vacuum deposition of silver or aluminium on to the moulded surface of the replica. Often the use of phase-contrast illumination will give good contrast on the unsilvered replica.

Should more rigid replicas seem desirable, a moulding resin of the epoxy type may be used. These have a slight disadvantage in taking longer to set before they can be removed. They may also be more difficult to prise free.

Fig. 102 shows a comparison between a field viewed directly with

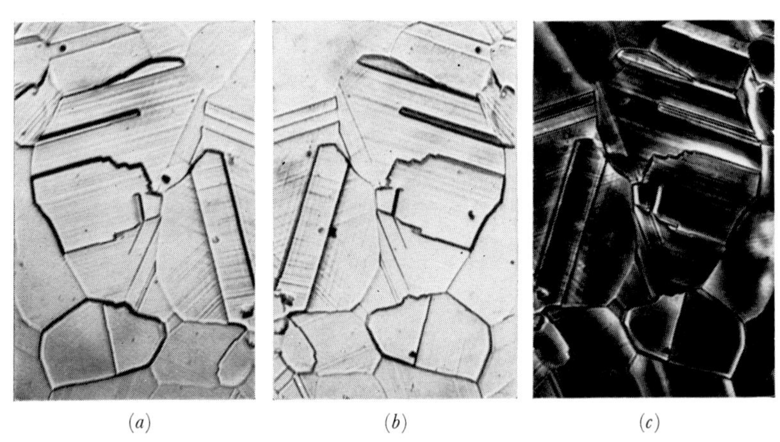

(*a*) (*b*) (*c*)

Fig. 102. Comparison of micrographs from
(*a*) a lightly deformed α-brass specimen;
(*b*) a metallized replica of (*a*), normal illumination;
(*c*) as for (*b*) but with negative phase-contrast. × 100.
(Reduced to $\frac{9}{10}$ in reproduction)

bright-field illumination and a replica of the same area viewed both with bright-field and phase-contrast illumination.

4. Stereomicroscopy

Many specimens, particularly those having fracture surfaces, damage or corrosion, are such that a three-dimensional view is of great assistance in studying them. Until very recently this could be realized only with *stereomicroscopes* which consisted of two complete sets of objectives and oculars (one set for each eye); moreover, for reasons which will be explained, the magnification with this type of microscope is limited to about × 200, but generally closer to × 100. It is possible now to obtain true stereoscopic vision using a single objective on a normal microscope by means of a device known as a pupil splitter; this system is available on one make of microscope (the Bausch and Lomb).

CONVENTIONAL STEREOMICROSCOPES

Fig. 103 (*a*) illustrates the principle of the conventional stereomicroscope. The three-dimensional appearance of the image depends upon each eye being presented with a view of the object from a slightly different angle; this, of course, is the basis of the stereoscopic vision of the unaided eye. Several factors limit the magnification available with this type of microscope. The objectives need to be close together, in order to keep the divergence of the two beams close to the natural convergence angle of the eyes at the point of nearest distinct vision, otherwise the two images will be of such widely differing aspects of the object that the brain will be unable to put them together to form a recognizable three-dimensional picture. The second point is that the depth of field must be large enough to give an acceptable third dimension to the image. In typical specimens the depth of field required is of the order of 10 μm. Reference to Table 1 (p. 44) will show that this requires an N.A. of about 0·20. This, in turn, leads to a primary magnification of about × 15, hence a total magnification (with × 10 oculars) of × 150. In practice, a third factor of importance tends to reduce this magnification. This is the desirability of having objectives of long working distance. Designing for this involves sacrificing magnification. A little of this can be regained by using an auxiliary magnifier between the objectives and the specimen, and so we finish with a magnification of about × 100. With special cases, using higher-powered oculars, a magnification of × 200 may be reached. A typical stereomicroscope is shown in Fig. 103 (*b*).

It is an advantage to have several sets of paired objectives mounted in a revolving nose piece or drum, so that magnification may be stepped up quickly on the same field. Many stereomicroscopes now have an additional facility whereby the magnification given by the oculars may be varied continuously over a range of about tenfold by rotating a telescopic component, known as a "zoom" attachment.

As in general work on macrostructures, correct lighting of the specimen is important to give satisfactory moulding of the structure. Small lamps which can be focused to give broad spotlight beams are usually employed; often they are mounted on jointed arms, so that the direction and elevation of the beam with respect to the specimen can be altered quickly and positively. A recent and highly satisfactory method of lighting employs several *fibre optic* illuminators mounted around the specimen stage. The principle of fibre optics is that of providing a thread of glass, or other transparent medium, such that internal reflections cause it to act as a waveguide for light; the light is "piped" from a source, through the fibre, to the point at which illumination is required. Such fibres, when fine enough, are flexible and guide the light even when bent. A bundle of fibres about $\frac{1}{8}$ in. in diameter will conduct sufficient light from a small bulb to illuminate a specimen for stereomicroscopy with an approximately parallel beam.

This kind of microscope may be used to take stereo pairs of photographs by exposing film through each arm of the microscope in turn. The

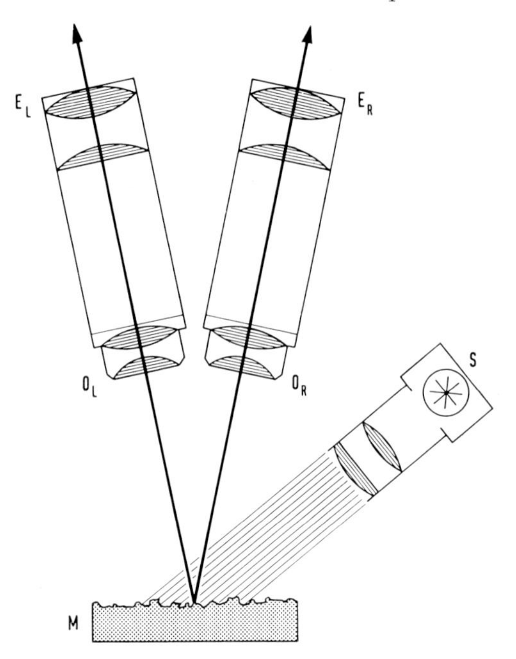

Fig. 103 (*a*). Principle of the stereomicroscope—two complete microscopes are used to present the right and left eye with slightly different views of the specimen.

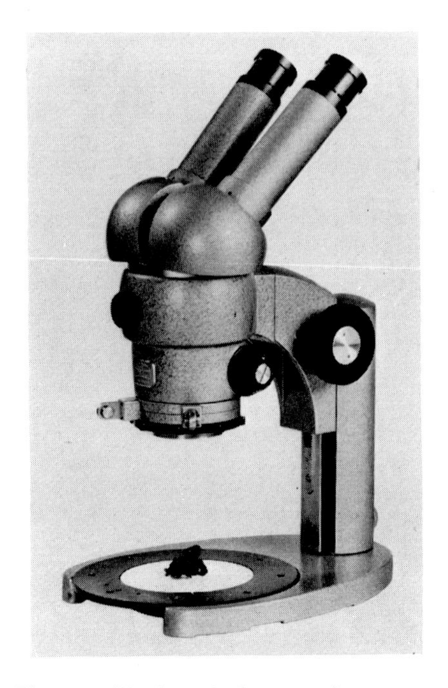

Fig. 103 (*b*). A typical stereomicroscope.

three-dimensional image is then reconstructed by mounting the pair of prints in an appropriate viewer which has a pair of low-power lenses. It is possible to obtain the three-dimensional effect by squinting at the stereo pair with a piece of card held between the photographs, normal to their plane. It is not easy to do this.

SINGLE-OBJECTIVE STEREOMICROSCOPY

The manner in which the ordinary single-objective microscope may be modified to give a true stereoscopic image (as distinct from mere binocular vision) is simple in principle.[9] It will be seen in Fig. 104 that if we imagine

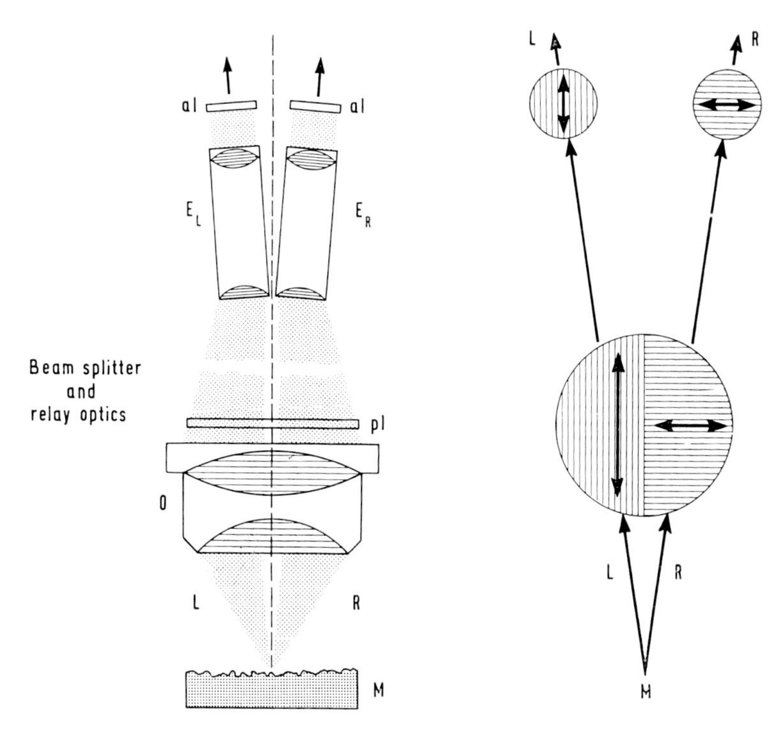

Fig. 104. A single-objective stereomicroscope. The two beams, one for each eye, are obtained by the use of a split polarizer (pl) and matched analysers L and R.

the objective to be in two halves, there are two beams L and R diverging from the specimen. We will here consider only the light after reflection from the specimen and ignore, for the moment, the problem of illuminating the specimen. If it could now be arranged that the left eye received only the L beam and the right eye only the R beam, it should be possible to see a true stereoscopic image. A complication which arises here is that this image would then seem to reverse elevations and depressions. This is overcome by arranging for the left eye to receive the R beam and the right eye the L beam. However, if the microscope incorporates optical arrangements which erect the image (i.e. render it right way up with respect to the specimen) this crossover is not necessary.

To achieve the separation of the L and R beams in this way they are passed through pieces of Polaroid sheet placed at the rear of the objective, at its exit pupil. The planes of polarization of the two beams are arranged at right angles (to within about 1°) as indicated in Fig. 104. Second pieces of Polaroid are used as analysers at the oculars, with their planes of polarization chosen to pass the light of the L beam to the right eye and that of the R beam to the left. In this way each eye sees the appropriate half beam only, and a stereoscopic image is constructed. The depth of field remains the same as for the particular objective used in the conventional manner.

In a metallurgical microscope the incident light passes through the exit pupil of the objective on its way to the specimen, so that the arrangement in Fig. 104 would result in stopping off the light on its return after reflection at the specimen. The split polarizer has, therefore, to be placed at a conjugate point, where the exit pupil is imaged beyond the beam-splitter. This is generally easy to do in large metallographs, because they incorporate an optical system to relay the light to eyepiece tubes or camera bellows placed some distance from the beam splitter. Some phase-contrast systems place the phase annulus at this conjugate point within the relay system. It is probably more difficult to put the polarizers in the equivalent plane in a bench microscope. The other practical difficulty is that of mating the two pieces of Polaroid together without overlapping or creation of a gap; this would appear to be a matter of accurate machining of their edges.

This system of stereoscopic microscopy may be employed with dark-field or phase-contrast illumination and at whatever magnification the objectives are capable of achieving. Just as with prism illumination, resolution is impaired because only half of the N.A. is realized.

5. A Universal Tilting Stage

In the examination of features such as fractures where surfaces of interest may be present at various angles to the general level, or where they are situated within cavities or crevices, it is convenient to be able to tilt the stage carrying the specimen quickly and positively about more than one axis. McNeil[10] has made such a stage, two views of which are shown in Fig. 105. The specimen is mounted on a disc which moves eccentrically over a second larger disc, and therefore provides a means of traversing the specimen in two orthogonal directions. These discs are carried on a tube which may be raised or lowered through a sleeve, and this assembly is capable of being both rotated and swung with respect to the frame. The frame itself is clamped to the microscope stage. With these movements it is possible to bring any facet of a specimen into a horizontal position. The working distances of normal objectives limit the magnification which can be used, but with a 2 mm reflecting objective McNeil reported that fracture surfaces had been successfully photographed at × 750.

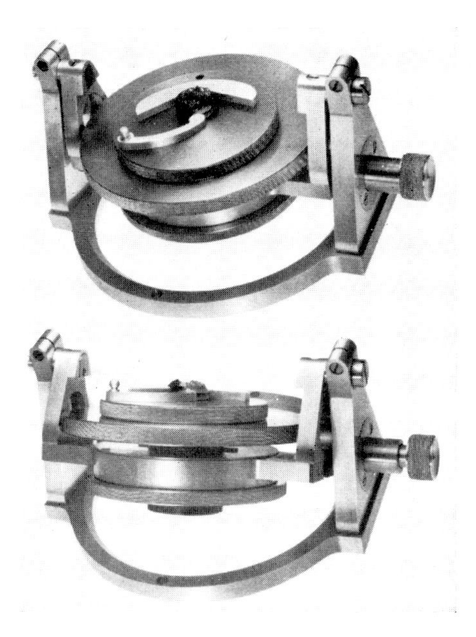

Fig. 105. A universal tilting stage for the examination of particular areas of multi-level specimens; the tilting allows any level to be brought into focus. (McNeil[10].)

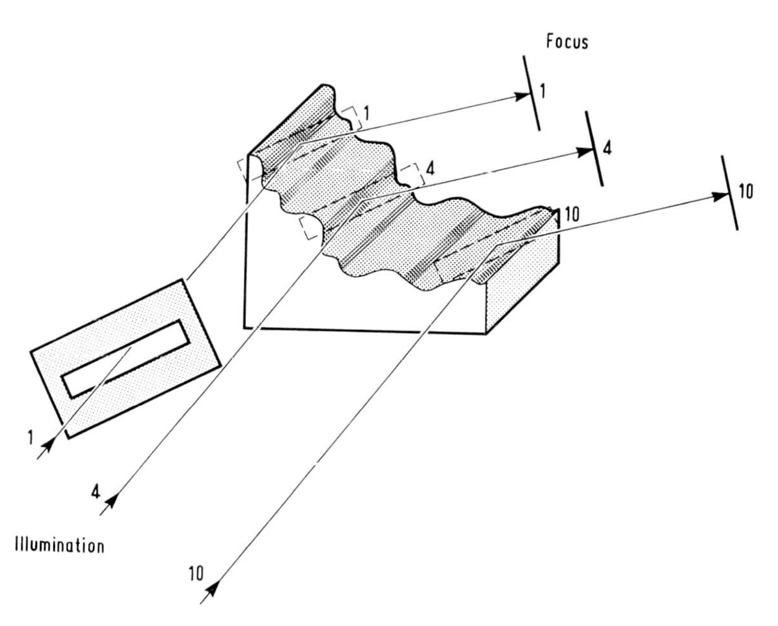

Fig. 106. Arrangement for obtaining photographs of "extreme focal depth". The specimen is illuminated strip by strip and the camera moved synchronously with the illumination, to build up a sharp photograph of the entire surface. (McLachan[11].)

6. Extreme Focal Depth

McLachan[11] has extended the Schmaltz technique (see Chapter 6), to enable an increased depth of focus to be obtained in photography of surfaces having levels too diverse for the inherent depth of field of the microscope to encompass. A strip of illumination 2 or 3 mm wide is arranged (Fig. 106) to fall across the level of the specimen which is in focus and this is recorded on a photographic plate. It is arranged that the strip of light moves as the focus is changed to adjacent levels and, in this way, a sharp photograph of the whole series of levels is built up. The diagram shows three of ten such positions in a series of exposures. The picture obtained is not three-dimensional in the usual sense, that is, it is not a perspective view but an isometric one. The detail revealed depends on the shadowing and obstruction effects which result from the directions of illumination and viewing of the specimen. It should be noted that this technique is still limited by the intrinsic depth of field of the microscope, the depth in each strip being unaffected.

7. Reflecting Objectives

It is possible to make a microscope objective using curved reflecting surfaces to bend the light rays and form an image, in contrast to the usual objectives in which refraction within the lenses occurs. Reflecting objectives have two theoretical advantages. The first is that they may be designed to have a much greater working distance than a refracting objective of equivalent power and N.A. As an example, a reflecting objective of N.A. = 0·95 may have a working distance of 5 mm compared to 1 mm for the equivalent refracting objective. The second advantage is that a reflecting objective brings all wavelengths of light to the same focus because there is no dispersion on reflection. Thus, full achromatic correction is automatically achieved and also the possibility exists to use ultra-violet (or infra-red) radiation and know it will be focused exactly as visible wavelengths. This latter property was the principal motive for developing reflecting objectives for biological work, but the long-working distance is the property which has generally been exploited in metallography.

There are two basic types of reflecting objective—the Schwarzschild (Fig. 107 (*a*)) and the Cassegrain or Newton (Fig. 107 (*b*)). In both types the rays diffracted by the object are reflected at two curved surfaces; in (*a*) they fall firstly on a concave mirror M_1 of large radius of curvature and then on a convex mirror M_2 of smaller radius of curvature; in type (*b*) the order of mirrors is reversed. In both arrangements the smaller central mirror stops off part of the cone of rays from the object (shown by the shaded areas in Fig. 107); this *occlusion* is less marked with Schwarzschild objectives, which is a reason for their generally being preferred.

The objectives may be used in the usual way in metallography, with the illumination provided by a beam-splitter. However, reflecting objectives are generally heavy and bulky and thus best suited for use on inverted microscopes.

Dyson[12] has designed an ingenious reflecting adjunct to a normal refracting objective, which confers a greatly increased working distance

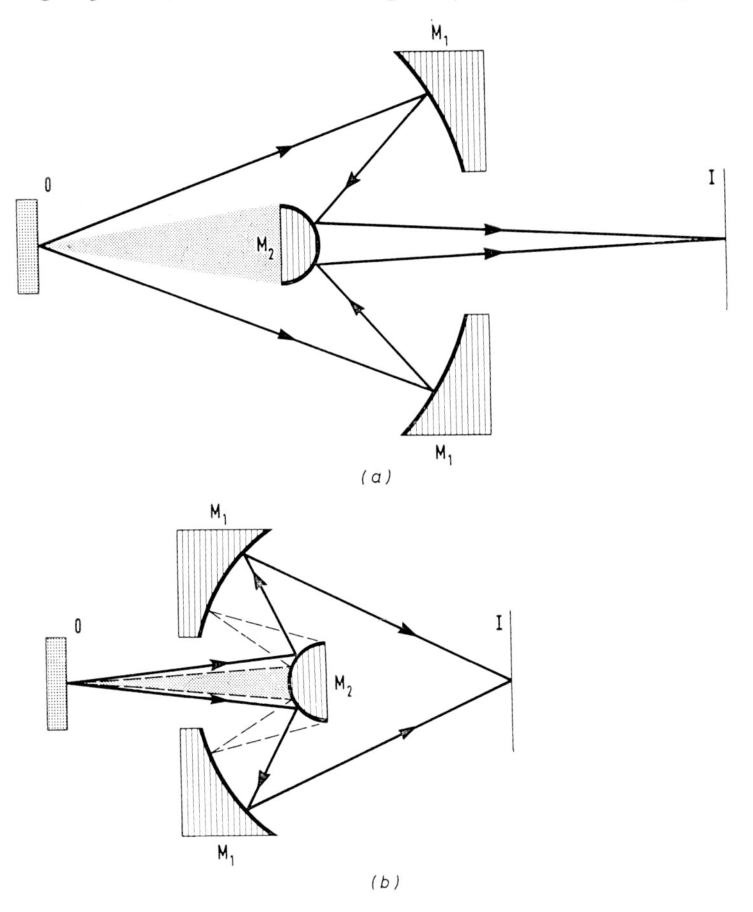

Fig. 107. Basic types of reflecting objectives.
(*a*) Schwarzschild.
(*b*) Cassegrain.
In each case two spherical mirrors are employed, one having a central gap to permit rays to reach the other.

upon the objective. This is illustrated in Fig. 108, in which the principle is self-evident. The working distance of a 4 mm objective is increased from about 1 to 13 mm with this attachment. This device is available commercially.

8. Hot-stage and Cold-stage Microscopy

A long working distance may sometimes be useful in examining fracture surfaces, particularly when used in conjunction with a tilting stage of the kind described on p. 156. However, by far the largest field of application for the reflecting objective has been in hot-stage and cold-stage metallography. When a specimen is examined whilst actually being held at an elevated temperature a long working distance is not only desirable to protect the objective from heat, but also may be necessary to allow room

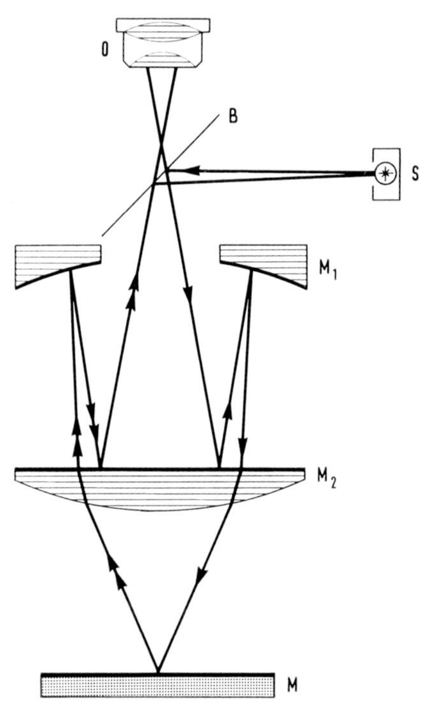

Fig. 108. Reflecting attachment which increases the normal working distance of an objective O. (Dyson[12].)

for the heating element, window, and vacuum chamber in which the specimen is enclosed. This chamber could also contain a gaseous mixture to promote a controlled reaction such as oxidation. In a cold-stage microscope the specimen is kept at some specified temperature below zero (0° C) by a suitable coolant, and the chamber and window are then necessary both to control heat ingress and to prevent condensation on the specimen. Often a jet of air is blown across the gap between the objective and the specimen chamber to assist in temperature control. The details of the mechanical and electrical requirement for a hot-stage are typified in the schematic diagram shown in Fig. 109.

Reflecting objectives with hot-stages have been used to study martensitic transformations, where the new phase causes a change in the surface contours which may be discerned without etching. The principal limitation would seem to be that the system of interest must show changes at the elevated temperature which are revealed without further treatment such as etching.

Whilst it has been implied above that hot-stage and cold-stage microscopy demand the use of a reflecting objective, this is only absolutely necessary when high magnifications and resolutions are required. If the demand is only for a modest magnification (say × 150) then a normal

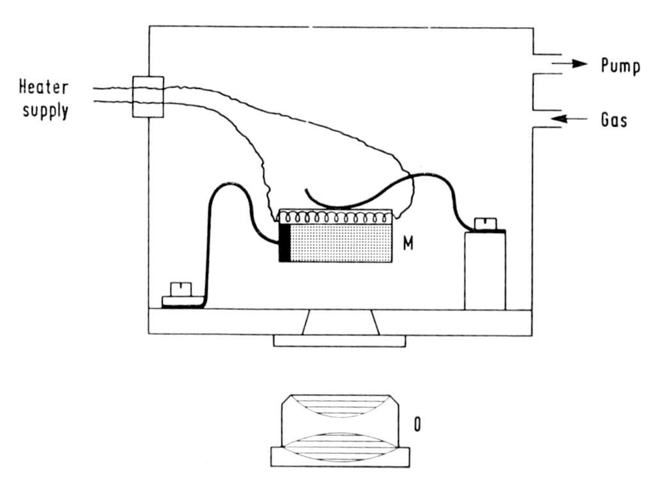

Fig. 109. Schematic arrangement for hot-stage microscopy; the specimen M is heated directly by a coil within an evacuable chamber and is viewed by the objective O through a window. (Brinson and Hargreaves[13].)

refracting objective may suffice and yet have sufficient working distance. Moreover, it can be used in conjunction with polarizing effects; in this way recrystallization of an anisotropic metal may be followed whilst at temperature. The oblique reflections which necessarily occur in a reflecting objective render it unsuitable for use with polarized light.

9. Image Projectors and Intensifiers

It is sometimes desirable to display the image formed by a microscope so that it may be viewed by several people. It is possible to do this by having specially designed projection oculars and screens, and Fig. 110 shows a metallograph with this feature. The image on this particular model may be viewed by a number of people provided they keep within a fairly limited range of angles about the normal to the screen.

Images may also be intensified and displayed by picking up the light with a small television camera which replaces the ocular of the microscope. After suitable amplification the image is then displayed on a standard television screen. This provides a very flexible tool for demonstrating metallurgical structures, the effects of illuminating conditions, and so forth, to large audiences. Contrast may be controlled through the electrical controls and a high degree of contrast enhancement employed. Resolution of detail is, of course, limited by the line spacing of the trace

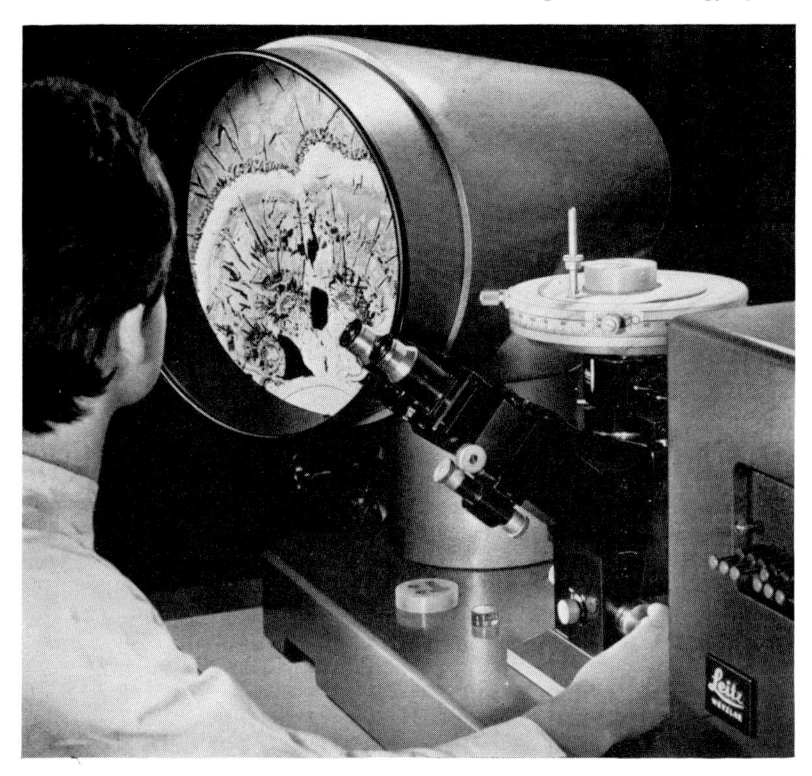

Fig. 110. Metallograph with image projector.

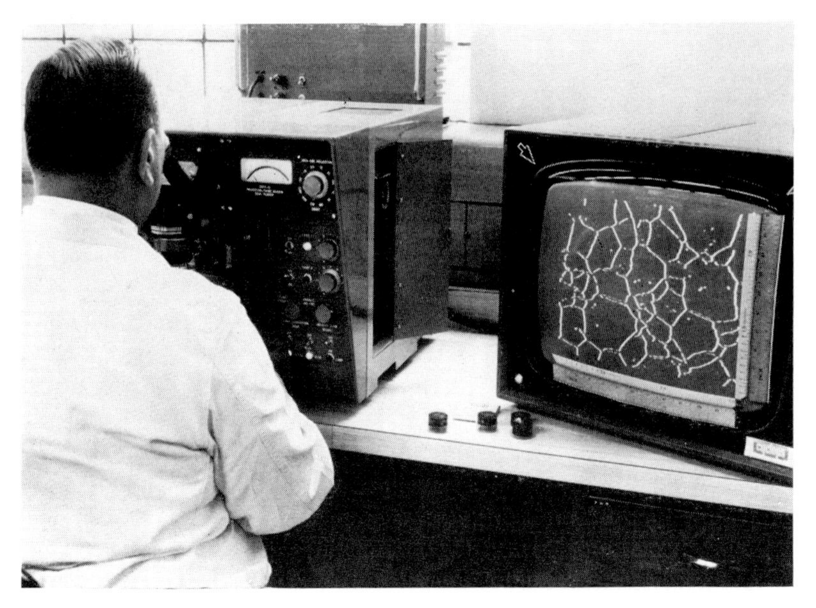

Fig. 111. Metallographic image displayed on a T.V. screen in an arrangement designed for automatic quantitative metallography. In the example shown, grain size is being measured.

on the television screen but, as always happens with resolution, is not a problem except in rather rare cases.

The other important and growing use for this kind of imaging device is in quantitative metallography (Fig. 111), where the contrast control is used to pick out specific features and the scanning aspect of the television camera leads directly to lineal or areal analysis (see Chapter 10).

"FLYING SPOT" MICROSCOPE

Roberts and Young[14] have described a method of image intensification in which a small spot on a cathode-ray tube is imaged on to the specimen, by placing the tube in front of the eyepiece of a microscope. The spot moves on to a raster, as in conventional television practice, and so scans the specimen. The light reflected from the specimen (actually Roberts and Young used the device in transmission) is projected on to a photomultiplier, and thus small changes in reflectivity may be amplified to give an image on a second cathode-ray tube. So far, the device does not appear to have been used in metallography.

10. Taper Sectioning

Although taper sectioning is perhaps best classified as a technique of preparation rather than one of examination, it is appropriate to include some notes on it here, because the information it yields may be so

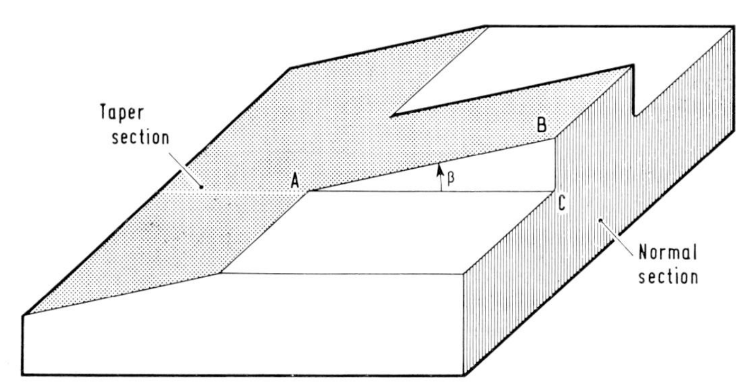

Fig. 112. Showing, in a simple case, how the taper section AB magnifies the step BC. The taper ratio is AB/BC = cosec β.

closely related to that obtained using interferometry and the light-profile microscope (Chapter 6).

Taper sectioning consists in cutting the specimen obliquely through the surface whose profile is of interest. Usually, of course, sections are cut normal to this surface and the profile is then seen with a magnification in depth equal to the lateral magnification employed. When an oblique section is examined, the profile is magnified further in depth by the geometrical effect of the taper section, as shown in Fig. 112 for a simple case.

Clearly, the step CB which is seen on the section taken normal to the surface becomes magnified to AB on the taper section, which is indicated by the dark shading. In this simple situation it is easy to see that the taper ratio or geometrical magnification due to the taper sectioning is AB/CB = cosec β, where β is the *taper angle* between the plane of sectioning and the original surface.

METHODS OF SECTIONING

Three methods of obtaining a desired taper ratio are in general use[15] and these are illustrated in the three sections of Fig. 113.

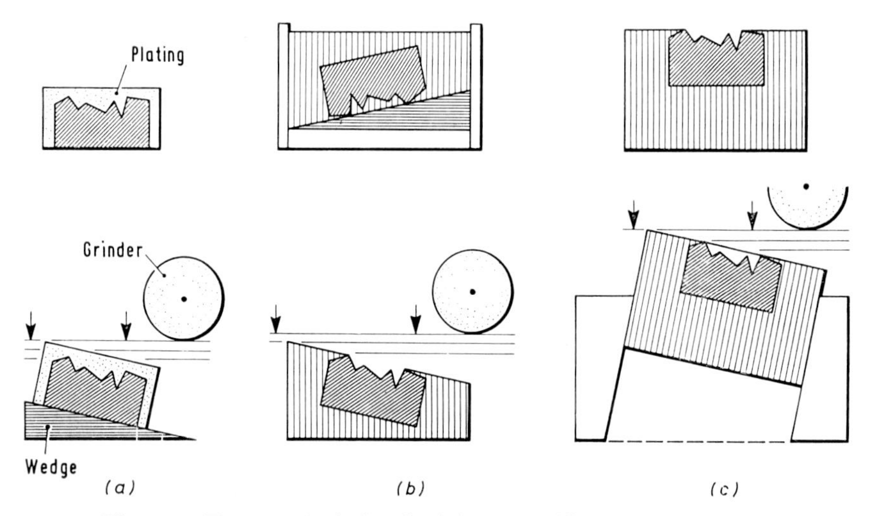

Fig. 113. Three methods for obtaining a specific taper ratio.
(*a*) Specimen set on a wedge and ground.
(*b*) Specimen mounted in plastic to hold it on a wedge, then ground.
(*c*) Specimen mounted in plastic and held in a block having a slanted hole, then ground.

In method (*a*), the specimen is held on a wedge-shaped block and ground as indicated in the diagram; the taper angle is set by the angle of the wedge. It is necessary to protect the surface of the specimen with a heavy layer of plating.

Method (*b*) uses a wedge to set the surface at the taper angle, whilst the specimen is being embedded in a plastic mounting material. The mounted specimen is then ground parallel to the base of the mount.

The third method (*c*) again involves mounting the specimen in plastic, but the mount must be of a standard size. This is in order that it shall fit tightly into a hole reamed at the required taper angle through a block. The block is split so that it may be tightened firmly on to the mount. The taper section is then made by grinding parallel to the base of the block.

Grinding tends to damage the specimens to an extent which causes difficulty in obtaining a good metallographic polish and using a surface cutter is preferable. With method (*c*) it is possible to mount the block on a lathe and section by turning, which is also preferable to grinding.

A special method of taper sectioning may be used with cylindrical specimens[16]. A flat is cut or ground parallel to the axis of the specimen and, in this way, a taper ratio established along the edges of the flat. The taper angle is determined by the angle between the flat and the tangent to the cylinder at the edge of the flat. However, the taper ratio varies as the field viewed is moved away from the edge, and this method is therefore less easy than the others to use quantitatively.

MEASUREMENT OF TAPER ANGLE

It is very usual for a taper ratio of 10:1 to be used; the taper angle β is then $5^\circ 43'$. It is not always easy to obtain such a small angle accurately, and relatively small errors have a large effect on the taper ratio. For example, an error of $3'$ will change the taper ratio by 1 per cent. It is therefore desirable to have some means of measuring the actual taper angle achieved, and not to trust the nominal value.

It is possible to obtain a measure of the taper angle directly using a method which determines the profile of the specimen (or specimen in its mount) as seen in the appropriate side elevation. Various profile projectors (of the kind mentioned in Section 2 of this chapter) are available to do this. However, it would not be worth while acquiring such a projector for the sole purpose of measuring taper angles. It must also be remembered that the profile should be measured after all mechanical preparation has been completed. This is because it is very difficult to maintain the taper ratio unchanged during mechanical polishing; some tilting and rocking readily occurs and is very likely to lead to an error of 10 per cent in the taper ratio.

A very satisfactory way of measuring taper ratio[15] is to place a piece of wire along the surface of interest, the wire being as closely as possible perpendicular to the line of sectioning. Fig. 112 will serve to illustrate this if the rectangular protrusion is imagined to be replaced by a wire of circular cross-section, with diameter equal to CB. It will be realized that the wire will present an elliptical section on the taper section, such that the major axis of the ellipse is equal to AB. Hence, the taper ratio is determined at the point B as the ratio of the major axis of the ellipse to the diameter of the wire. The latter should have been determined accurately and the wire checked for uniformity before mounting it with the specimen in the manner to be described.

The specimen is mounted in plastic and two slots made one either side of the specimen. The wire is bent and pushed into these slots, so that it is held both close and parallel to the surface of the specimen. The wire is prevented from moving during machining by being embedded in a thick layer of electrodeposit, chosen so that it adheres both to the wire and to the specimen. Care must be exercised, of course, to ensure that the process of electroplating does not corrode or damage the surface to be examined. Wire of 35 s.w.g. is suitable for a taper ratio of 10:1.

12—(T.220/M.6601)

APPLICATIONS

The total vertical magnification is greater the greater the lateral magnification at which the taper section is viewed; this means that the limitations on magnification and resolution are those which normally apply. Hence, also, taper sectioning to examine profiles has an advantage over interferometric methods, because the highest magnifications using oil-immersion objectives may be used whereas interferometry is limited to relatively low numerical apertures and magnifications of less than × 500. However, the light-profile techniques may also be used with the highest magnifications.

It should also be noted that there are profile instruments which use some kind of probe to sense the profile and which will give vertical

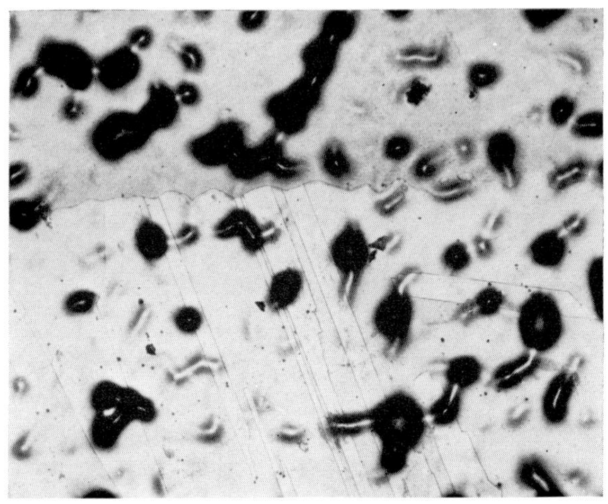

Fig. 114. Artefacts in a taper section, caused by irregularities (the dark areas) in the protective electroplated layer. Lateral magnification × 100; vertical magnification × 1000.
(Reduced to $\frac{3}{4}$ in reproduction)

magnifications of at least × 500, but very low lateral magnifications. The probe may be a diamond stylus or a small jet of compressed air, and the magnification is obtained by mechanical means.

In general, it would seem that taper sectioning has advantages where one or more of the following conditions pertain. First, if the vertical magnification desired is larger than a few hundred times, particularly if this is linked with the necessity for having a large lateral magnification (but note that these conditions are met by use of the light-profile microscope). Second, if the surface is not sufficiently smooth or reflecting to form good interference fringes (but note that this may be overcome sometimes by depositing silver or aluminium on the surface). Third, if the features of interest are too small and too narrow (e.g. fine fissures or cracks) for light to be reflected back into the objective from within them or for a probe to penetrate them; in this case taper sectioning has unique advantages (including those listed in the fifth case below). Fourth, if the surface is too rough either on a coarse scale (e.g. in fracture surfaces) when problems of depth of field arise with normal optical methods, or on a fine scale (e.g.

heavily deformed or etched specimens) when problems of low reflectivity again arise, taper sectioning has clear advantages here. The fifth case, for which taper sectioning has unique advantages, is not concerned with profiles but with the structures of thin layers. The surface damage caused by mechanical treatments such as grinding, by oxidation or by nitriding are typical of this kind of specimen. Several good examples will be found among the illustrations to Samuels' book.

We conclude with a word of caution on the interpretation of profiles revealed by taper sectioning. There is little difficulty in measuring uncomplicated ridges or grooves of the kind shown in Fig. 112, especially when the line of the section is symmetrically placed with respect to it. However, if such features are superimposed upon other changes of level, which present lines of greatest slope or greatest depth in different directions, the local taper ratio may not be that obtained from a straightforward calibration (see Fig. 114). Most people find it difficult to think clearly about three-dimensional objects and their sections (see Chapter 10) and it may assist in preventing mistakes to model the tentative interpretation in modelling clay.

REFERENCES

1 T. Berglund, *Metallographer's Handbook of Etching*, London, 1931 (Pitman).
2 *Metals Handbook* (1948 edn), Cleveland, (Amer. Soc. Metals).
3 C. S. Smithell, *Metals Reference Book* (4th edn), London, 1967 (Butterworth).
4 L. E. Samuels, *Metallographic Polishing by Mechanical Methods*, London, 1967 (Pitman).
5 P. Lacombe in *Metallography 1963* (*Proceedings of the Sorby Centenary Meeting*), London, 1964 (Iron & Steel Institute).
6 D. McCutcheon and W. Pahl, *Metal Prog.*, 1949, **56**, 674.
7 J. D. Livingston, *J. Aust. Inst. Metals*, 1963, **8**, 15.
8 D. H. Kay (ed.), *Techniques in Electron Microscopy*, Oxford, 1965 (2nd edn) (Blackwell).
9 H. E. Rosenberger, *High Power Stereo*, 1967 (Bausch and Lomb).
10 J. F. McNeil, *Metallurgia*, 1956, **54**, 207.
11 D. McLachan, Jr., *App. Optics*, 1964, **3**, 1009.
12 J. Dyson, *Nature*, 1949, **163**, 400.
13 G. Brinson and M. E. Hargreaves, *J. Inst. Metals*, 1957–58, **87**, 112.
14 F. Roberts and J. Z. Young, *Nature*, 1951, **167**, 231.
15 L. E. Samuels, *Metallurgia*, 1955, **51**, 161.
16 W. A. Wood, *Phil. Mag.*, 1958, **3**, 692.

10 Quantitative Metallography

Traditionally, metallography has been a qualitative method of investigation and a great deal of useful information has been obtained and catalogued with such qualifying words as "coarse", "fine", "massive", and so forth. The principal exception to this, until recently, has perhaps been the measurement of grain size, but even this has usually been done without attempting to relate the results to a three-dimensional situation.

This, of course, is the point which must always be borne in mind; the microstructure seen is a two-dimensional view of a three-dimensional structure. C. S. Smith has done more than anyone else to put this aspect of metallography on a systematic and sure foundation, and his review papers on the subject are recommended reading.[1,2] It is not possible within the scope of a chapter to cover the whole range of the topic of quantitative metallography, but it is so important that a book of this kind would be seriously at fault if it neglected to call attention to some of the more important ideas on the subject.

It should be noted that the difficulties associated with studying solids through the appearance of sections are not peculiar to metallography; botanical, biological and geological systems all have similar distributions of phases, cells and membranes. Thus, many of the detailed relations of use in metallography have been specifically derived for non-metallurgical systems, and the general title "quantitative stereometry" has been coined to cover this field of study. An excellent general treatise by Dettoff and Rhines[3] has appeared recently.

What we require, therefore, is firstly an understanding of how space can be filled by various grains and phases, based on the physical and geometrical laws which govern their mutual adjustments of shape. Secondly, we need rules for converting the measurements that can be made on polished sections into useful parameters concerned with shape and distribution. As may be imagined, there are a great many rules concerned with specific shapes, some of which may be rather rarely found. Hence an attempt will be made here to review, rather than develop, the ground rules of the subject. Since a good deal of interpretation of quantitative data of any kind turns upon statistical treatment, an appendix on simple statistics has been included. The concepts and rules given there should be adequate for most everyday cases of quantitative metallography, but it might be more prudent to regard it as either a revision of or a lead into the study of more extended and rigorous treatments.

Filling Space

The most restrictive conditions upon the shapes which will fill space completely without voids arise in connection with "equiaxed" grains of single-phase alloys. This is important, of course, in the most commonly made measurement in quantitative metallography, namely, grain size. If the grains are assumed to be of identical shape and size, they must be polyhedra, with plane faces, of one of four kinds. These are cubes, hexagonal prisms, rhombic dodecahedra, and truncated octohedra (or tetrakaidecahedra). The last two are the nearest to reality, because they present essentially the correct shape when sectioned randomly—see Fig. 115

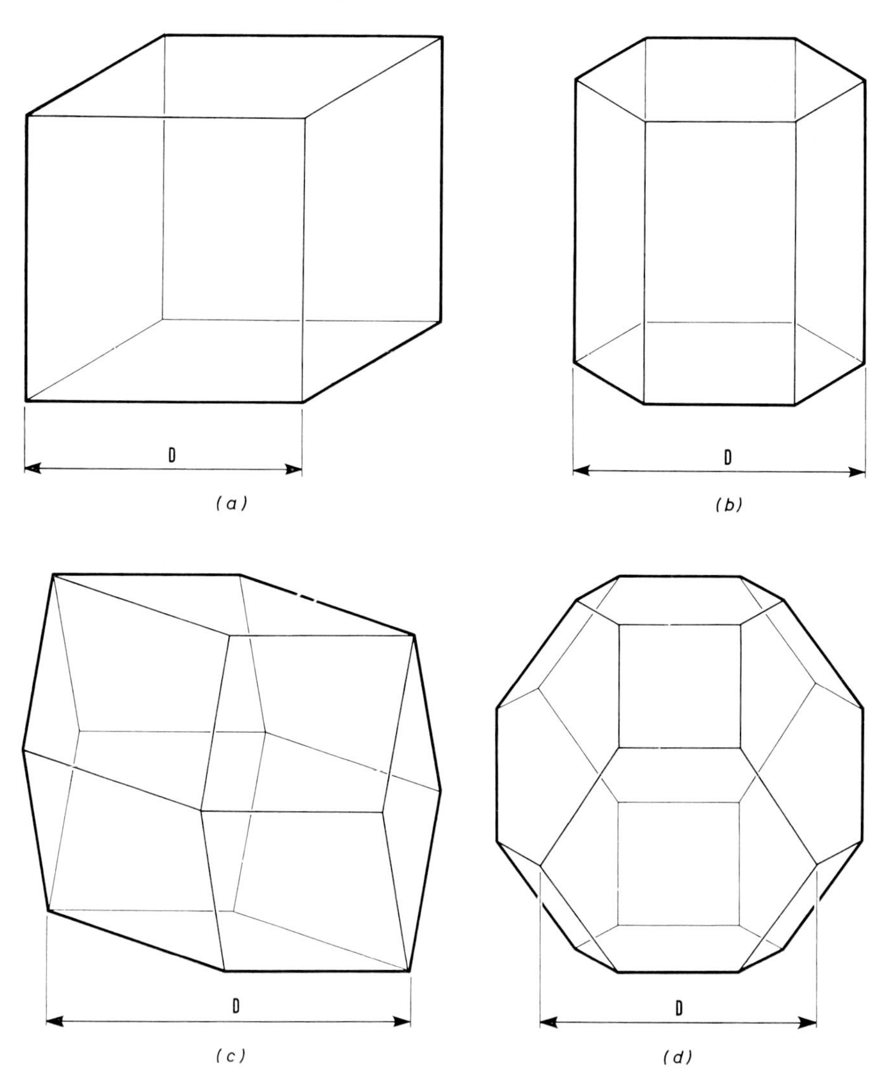

Fig. 115. Space-filling polyhedra which could be taken to represent idealized grains; (*a*) cube; (*b*) hexagonal prism; (*c*) rhombic dodecahedron; (*d*) tetrakaidecahedron (or truncated octahedron).

Clearly the other polyhedra only show no better than approximately the correct sectional view when cut in some specific ways and obviously incorrect sections in some other directions (parallel to the prism faces in the hexagonal prism, for example). It can also be shown that the tetrakaidecahedron has the advantage of allowing boundaries between grains to meet at the correct dihedral angle of 120°, with only slight curvature to the planar nature of the faces; this particular solid is then known as the Kelvin tetrakaidecahedron. A much favoured assumption is that grains are approximately spherical. It is easy to see that this must be approximate, because space is not filled by spheres; their packing results in the formation of voids. However, the error in computing some parameters to characterize grain size on this model may not be too great. The diameter of the appropriate sphere, for example, is not very different from the diameter of the tetrakaidecahedron measured between the square faces. We shall return to this point when the measurement of grain size is discussed separately in greater detail.

If the restrictions upon the size and shape of the phase are relaxed so that a range of sizes and more than one geometrical shape are permitted, the possibilities of filling space with the phase become infinite. Even so, it is possible to infer a great deal about the shape and distribution of such a phase from properly arranged measurements. Similar remarks apply to mixtures of phases. Some of the basic relationships pertinent to such observations will now be described.

Methods of Measuring

If it is required to gather quantitative information from the pattern revealed on a polished micro-section, what kinds of measurement may be made? To help answer this question a simple graphical analogue to a specimen will be used in Fig. 116. Here the polished section reveals a series of squares of a second phase and the "micrograph" has had a uniform grid superimposed on it.

AREAL ANALYSIS

An obvious way of saying something about this "square" phase is to measure the proportion of a given area of section which is occupied by it. This is known as *areal analysis*. In Fig. 116 it is comparatively easy to do this because the phase always appears as a square; it would only be necessary to measure the edges of a sufficiently large sample of the phase to obtain an accurate estimate of its total area relative to the section. If the phase were not square but irregular in shape, its area could be found by counting squares of the grid which fall within the phase boundaries. To obtain a reasonable accuracy and obviate the necessity for estimating fractions of grid squares, the grid would need to be much finer than in Fig. 116. This would be a tedious but otherwise practicable method. Another technique, much used at one time, was to cut out all the areas of the second phase from a photograph and weigh these pieces, comparing their total weight with the original weight of the uncut photograph. An instrument designed for map reading, and known as a planimeter, may also be used to assess areas if they are not too irregular. The planimeter has a point which is used to trace around the periphery of the area, and

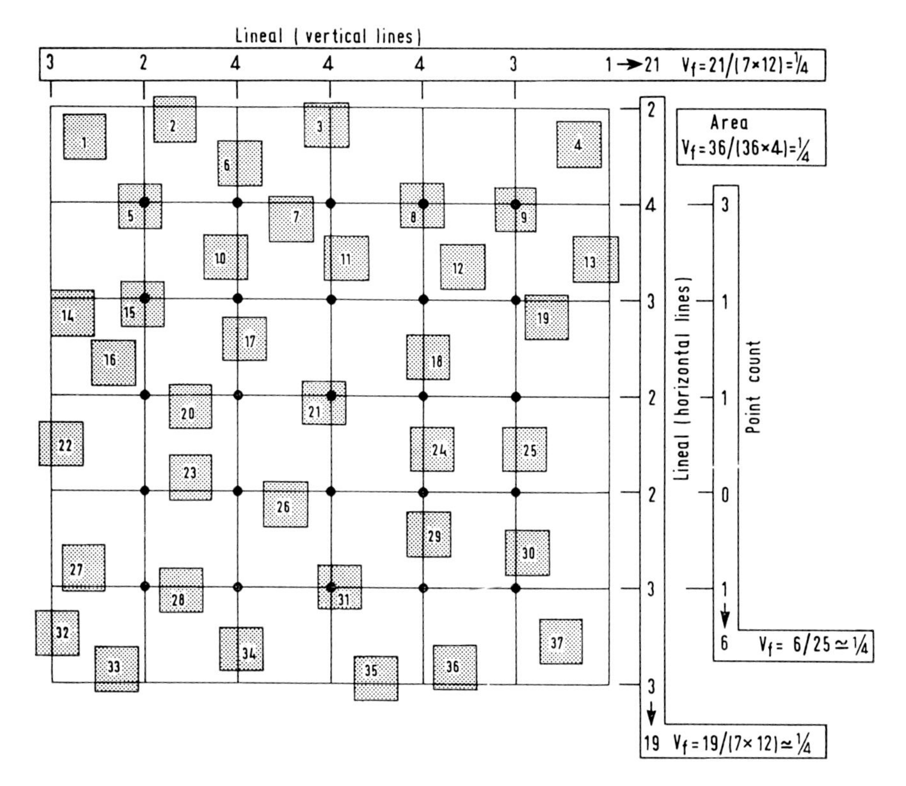

Fig. 116. Illustrating various methods of analysing a structure revealed by sectioning and polishing. The grey squares represent a second phase. The grid of lines is used to make areal, lineal or point-counting analyses, which are recorded in the figure. The equivalence of the three methods in determining volume fraction can be seen.

a screw thread integrates this lineal traverse into an equivalent regular area.

Whichever method is used, let the area of the square phase measured be A_p in a total area of A_T; the ratio of these quantities is an important quantity which will be symbolized by V_{fa}, i.e.

$$V_{fa} = A_p/A_T \tag{1}$$

LINEAL ANALYSIS

The second way of measuring some attribute of the square phase relative to the matrix is to measure the proportion of a grid line which lies over each phase. This is much easier to do on Fig. 116—by counting units along lines—than it is to measure areas by counting grid squares. It could also be done quite readily by making the measurement with a scale in the microscope eyepiece.

Rather less readily, it could also be measured using the vernier scale on a microscope stage, noting readings as the boundaries of the square phase particles moved past a cross-wire in the eyepiece. This method may be mechanized quite easily; a constant rate of traverse of the stage is used to sense distances along the grid line and these distances are put into groups or classes by depressing appropriate keys. A successful commercial device of this kind is the Hurlburt mechanical counter. Other devices may be completely automatic, eliminating even the necessity for keys to be depressed manually. The principle used is for a photoelectric cell or T.V. camera to scan the specimen and for appropriate circuitry to be arranged to register phases of chosen contrast or depth of etch. Some difficulties may arise from lack of discrimination by this kind of instrument; it is not always possible to obtain etching which is clean and of sufficient contrast to differentiate the phases or particles of interest. On the other hand, accidental features such as pits, or even heavily etched grain boundaries, might sometimes be erroneously counted. These problems are such that they can be overcome by increasing the complexity (and cost) of the equipment and modern devices using T.V. cameras are both flexible and reliable, but expensive. Automatic counting of this kind is almost essential if enough readings to give high statistical significance are to be taken—the work of weeks done manually may be completed in minutes. As for areal analysis, the ratio of lineal intercepts L_p and L_T for two phases is important, i.e.

$$V_{fl} = L_p/L_T \tag{2}$$

POINT COUNTING

Another feature of Fig. 116 which gives information about the two phases is a count of the number of grid intersections which fall above each phase. With a regular grid it is only necessary to count the intersections above the lesser areas of the square phase, since the total number of intersections is known. It is rather easier to make the count if the grid intersections are emphasized by small crosses; the centres of the crosses then become approximately points and the method of measurement is known as point counting.

Point counting may be done very conveniently using an appropriate graticule (or reticle) in the eyepiece of the microscope. It has been shown[3a] that the best accuracy is obtained with a regular rather than with a random array of points as the reference pattern. This is fortunate, for visual fatigue is diminished by having the points placed so that the eye can scan them systematically. It is also an advantage to keep the total number of reference points in the test pattern to about twenty. One grid of points which is available commercially on a graticule and which meets these conditions is shown in Fig. 117. The procedure for using such a grid is very simple. The number of points lying over the phase being measured is noted, a score being made each time a test point falls within the phase. The total score for each application of the test pattern is noted and the

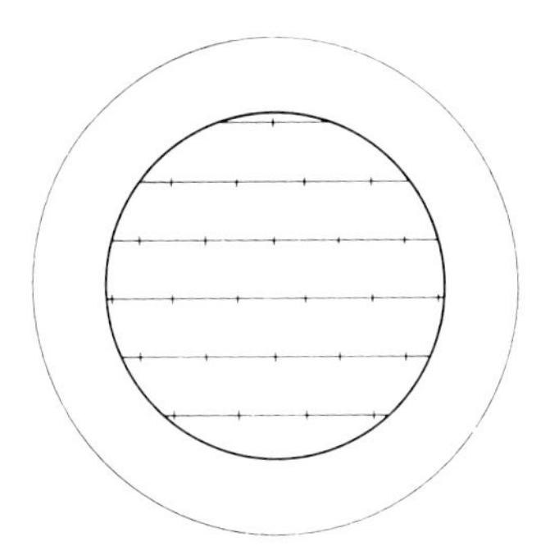

Fig. 117. Reticle for point counting using a microscope ocular (this particular design is that used in the Zeiss "Integrating Eyepiece").

means of these values P_P calculated. Once again (with P_T as the total number of points) the ratio of points is defined as

$$V_{fp} = P_P/P_T \tag{3}$$

VOLUME FRACTION

The importance of the ratios given in equations (1)–(3) is that

$$V_{fa} = V_{fl} = V_{fp} = V_f$$

V_f is the volume fraction of the second phase. This can be written

$$V_f = V_P/V_T = A_P/T_T = P_P/P_T \tag{4}$$

It can be shown that equation (4) is true whatever the shape of the second phase. It can readily be seen by actual counting that it is certainly true if the squares seen in Fig. 116 are sections of cubes. In that case we have

$$A_P = 36 \text{ and } A_T = 144$$

$$L_P = 21 \text{ and } L_T = 84$$

$$P_P = 6 \text{ and } P_T = 25$$

so that $V_f \simeq 1/4$. It would perhaps be less easy to see the truth of the relation if the cubes had been randomly oriented, so that their sections appeared on the polished surfaces as rectangles and triangles (Fig. 118). However, with the aid of the three-dimensional view (normally denied to the metallographer but also shown in Fig. 118) it is again possible by counting to verify equation (4).

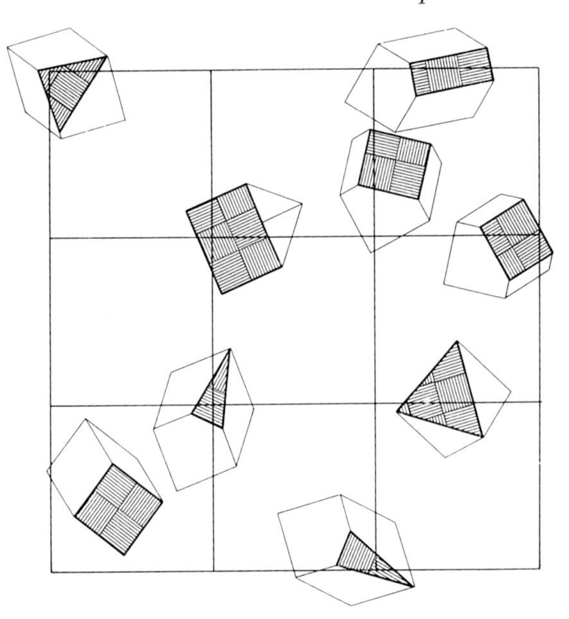

Fig. 118. As for Fig. 116, but having the grey (cubic) phase distributed randomly; the equivalence of the various methods of determining volume fraction also holds in this case.

SPACING OF PHASES (PARTICLES)

One other feature of the structure in Fig. 116 which may be important is the spacing between the areas of the square phase. This is often of great importance when the second phase is a finely dispersed precipitate. There are various ways by which the spacing could be characterized. On a section such as shown in Fig. 116 the spacing might be taken as the mean of the distances between the adjoining pieces of the square phase, measured along a particular grid line. This would give the planar mean-free path. On the other hand, should the phase be less symmetrically shaped and more randomly oriented within the matrix, the important parameter would be the mean distance between the nearest points of the phase, the volume mean-free path. The spacings revealed in a section would not, in general, be representative of these distances of closest approach. Fortunately, it is possible to obtain a general expression which indicates the volume mean-free path, M_F. It is

$$M_F = (1 - V_f)/N_L \tag{5}$$

where N_L is the number of particles intersected by unit length of a linear figure.

M_F, as the above discussion has indicated, is the mean distance from edge to edge of the particles. The quantity S_L, the reciprocal of N_L, is sometimes a useful measure of spacing, for it represents the mean centre-to-centre distance of the particles. When the particle size becomes very small, $(1 - V_f)$ approaches unity and thus equation (5) becomes

$$M_F = 1/N_L = S_L \tag{6}$$

Under these conditions, there is no point in making the rather more elaborate analysis to obtain M_F by equation (5); the simple count of particles per unit length gives S_L, and then M_F directly from equation (6).

Random Sampling

So far, the methods of sampling described have relied upon making measurements along one or perhaps two directions in the plane of the section; this has appeared to be a valid procedure because the example chosen in Fig. 116 is a simple case. Real structures in alloys tend to be very much less homogeneous and to have various kinds of preferred distributions of shape. Inhomogeneity can be on scales which vary from the sub-microscopic (which, by definition, need not worry the optical microscopist) to the grossly macroscopic. If the sampling cannot be controlled by qualitative selection, some scheme of random sampling must be used; in addition, of course, the measurements must always be processed using an appropriate statistical procedure.

It is not easy to select areas randomly by any conscious process. A method often recommended is to select a starting field, measure on it and then, closing the eyes, move the specimen by an undetermined amount to a second field and so on, but ensuring that this cannot lead to measuring on the same field more than once. This method will be successful in most cases. However, observers do tend to have unconscious patterns which they impose upon the controls moving the specimen. It could happen that these movements might bring banded or segregated areas into view, the pattern of movement more or less coinciding with the pattern of distribution. This is not, perhaps, very likely; but it should be borne in mind and an effort made to break any pattern. Making selections of numbered areas according to random numbers is usually possible. Taking the last digits from successive numbers in a column of log tables is one way of obtaining reasonably random numbers. Tables of truly random numbers are, in fact, available.

The other aspect of sampling concerns the direction of any line of traverse used in measurements. This should also be random. If a linear intercept or spacing is being determined, the line should be applied at angular spacings of about 15°, but otherwise the field is selected randomly. An excellent way to achieve this, due to Hilliard,[4] is to use a circle as the test figure. This effectively samples the structure along all directions. This device was developed for the measurement of grain size and will be discussed more fully in the section dealing with that subject.

Three-dimensional Shape

The examples discussed up to this point have been rather simple, and it has not been difficult to infer the real appearance of the solid phase or particle from the two-dimensional evidence, because we have assumed our section to be typical, whatever the sectioning plane. In general, of course, three orthogonal sections are required to ensure that the shape of particles is known, but two of these often suffice, if the correct pair is chosen. The most often quoted example of this is that of manganese sulphide inclusions in rolled steel bar or strip. Sections containing the longitudinal rolling direction give sausage shapes for the inclusions, but whether the inclusions are actually sausage-shaped (Fig. 119 (*a*)) or more

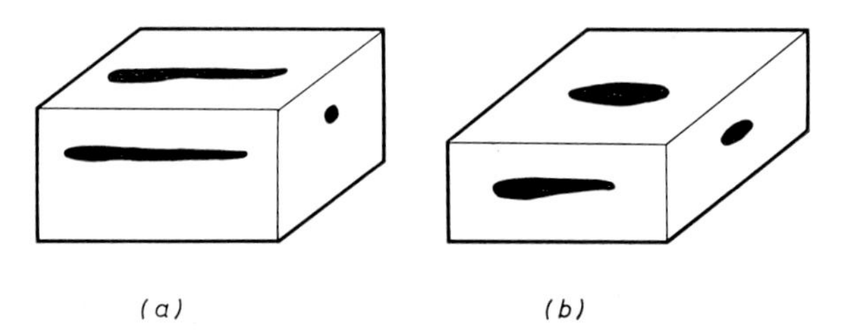

Fig. 119. Three orthogonal sections are required to reveal the true shape of non-metallic inclusions in rolled bar.
(*a*) Rolling mainly unidirectional.
(*b*) Significant cross-rolling.

disc-shaped (Fig. 119 (*b*)) depends upon the degree of cross-rolling; this may be gauged by examining the cross-section. Circular sections to the inclusion then imply sausage-shaped, oval sections rather than disc-like inclusions.

All this might seem trite and obvious, because we know a steel bar has certainly been rolled, and probably been cross-rolled, and this alerts (or should alert) us to the need for caution. Despite this, instances of persistent error are not difficult to find. For example, the fact that a polished and etched section of a martensitic structure gives an overwhelming impression of bunches of needles has earned the adjective "acicular" for martensitic

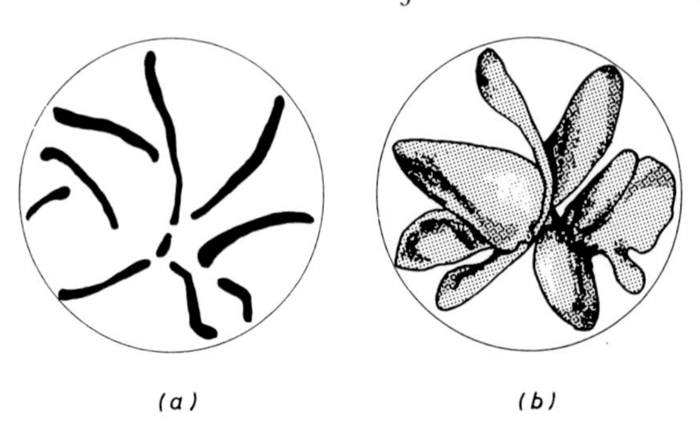

Fig. 120. Relation of graphite flakes (*a*) as seen in section to (*b*) true three-dimensional form (after E. E. Underwood[10]).

structures. The martensite is not, in fact, "needle-shaped" but "plate-like", although it is not easy to prepare sections which reveal this directly.

Another complex phase is that of graphite in cast iron. The graphite appears on a section as disconnected but related groups of "flakes" (Fig. 120 (*a*)) and it has not always been realized that these are sections through a continuous rosette of graphite (Fig. 120 (*b*)). This is probably the most complex three-dimensional pattern found in metals; fortunately, the majority of precipitates and transformation products belong to simpler systems—plates, rods, discs, spheres, ellipsoids, etc. Once the habit of thought has been reached, such that the question "What would it look like in the solid?" is automatically asked when a sectioned view is seen, it is generally relatively simple to answer the question unequivocally.*

Uniform Particles

It is a reasonable approximation for many alloys to assume that precipitate particles are of uniform size. Various measurements which may be made on a polished section may then be related to the true dimensions of the particles. Fullman[5] has worked these out for four simple shapes—spheres, discs, cylinders and rods. Table 5 sets out a number of parameters

TABLE 5

Numerical Relationships for Uniform Particles

Shape	$\bar{A}$	$\bar{L}$	N_V	V_f	Dimensions
Sphere r	$\frac{2}{3}\pi r$	$\frac{4}{3}r$	$\pi N_A^2/4N_L$	$8N_L^2/3\pi N_A$	$r = 2N_L/\pi N_A$
Disc r, t $(r \gg t)$	$2rt$	$2t$	$2N_A^2/\pi N_L$	$2N_L t$	$r = N_L/N_A$ $t = V_f/2N_L$
Cylinder r, h $(r \simeq h)$	$(2\pi r^2 h)$ $\div(\pi r + h)$	$(2rh)$ $\div(r + h)$	$(2N_A)$ $\div(\pi r + h)$	$(N_L 2rh)$ $\div(r + h)$	$r = (V_f/\pi N_V h)^{\frac{1}{2}}$ $h = V_f/\pi N_V r^2$
Rod r, l $(l \gg r)$	$2\pi r^2$	$2r$	$V_f/\pi r^2 l$	$2N_L^2/\pi N_A$	$r = N_L/\pi N_A$ $l = 2N_A/N_V$

connected with these types of particles. The following symbols are used in the table:

r = radius;
t = thickness;
h = height;
l = length;
$\bar{A}$ = mean area intersected by a random plane;
$\bar{L}$ = mean length intersected by a random plane;

* Since the above was written the increasing use of the scanning electron microscope (see p. 197) has helped to answer this question very clearly in many complex situations.

N_A = number of particles intercepted by a random plane;
N_L = number of particles intercepted by unit length of a random line;
N_V = number of particles per unit volume;
V_f = volume fraction of particles.

Inclusion Counting

A very important procedure in industry is that of characterizing alloys, particularly steels, by counting the number of inclusions in a standard area. The relevant A.S.T.M. Standard for inclusions in steels[6] gives considerable emphasis to qualitative methods, but the advent of reliable automatic counting devices probably means that any future standard will ask for quantitative results. The problem is, of course, quite standard and does not involve any principles or procedures other than those described in general terms for lineal analysis or particle counting. The reasons for giving it the prominence of a separate sub-section are that it is very important and that it happens to be very suitable for applying the type of counting device which uses a T.V. camera. The contrast between inclusions and unetched background is high, and therefore the instrument can be set to count inclusions alone, without occasionally counting some other features by mistake. It would seem to be important to gain general acceptance of the idea that a controlled count should be used to characterize the "dirtiness" of steel; the matter is too important to be left to subjective comparisons.

Grain Size

By far the most important metallographic feature which is treated quantitatively is grain size. There are many ways of determining and characterizing grain size. Too often metallurgists fail to distinguish their measurements by any indication of how they were obtained or, what is perhaps worse, use the wrong kind of parameter for the problem in hand.

As discussed in general terms earlier in this chapter, a good deal of the significance of grain-size measurements turns upon the assumed shape of the grains. There is little doubt that a tetrakaidecahedron is the space-filling regular polyhedron closest to the average shape of uniform grains. It fills space, allows the correct dihedral angle of 120° between grain faces to arise naturally and also gives approximately the correct distribution of polyhedral sections. This last point is worth elaborating. A cursory examination of a polished section of a polycrystalline metal or alloy might lead to the conclusion that most grains had 4–6 sides. This, in essence, is true. Some simple quantitative metallography directed to counting the number of sides per grain (on the section) would show that the mean value is about 5, but the range would be from 3 to about 8.

It will be seen by reference to Fig. 121 that there are many ways of characterizing grains of tetrakaidecahedral shape. The "diameter" could be defined in, perhaps, two ways. It could be the caliper diameter D_M;

this measures the maximum distance which could be gauged with a pair of calipers on the outside of the solid grain. With some justification, it could also be the maximum diameter of any face D_F; this measure appears to have been used occasionally and represents the maximum distance between grains measured along a grain boundary. On the other hand, D_M measures the maximum distance between grains measured

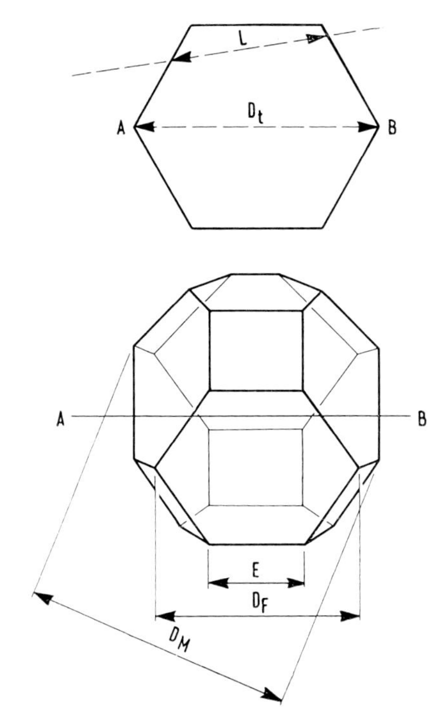

Fig. 121. Various measures of grain size referred to identical grains of tetrakaidecahedral form.
D_M = caliper diameter of grain;
D_t = diameter of grain section (here taken along AB);
L = lineal intercept on grain section;
D_F = diameter of (hexagonal) grain face;
E = grain edge.

through the grain. At other times the mean distance between grains measured either along grain boundaries or across the grains might be of more importance. Another parameter which is often relevant is the proportion of grain-boundary area to the total volume; this is an important case which will be dealt with later. There may be other occasions when the average volume of a grain is of interest. A tetrakaidecahedron has 14 faces, alternately squares and hexagons, there being 6 squares. The actual observed shape of real grains is perhaps more closely represented by the pentagonal dodecahedron, but uniform solids of this type do not fill space.

It is not always possible to measure all the above characteristics directly, even if the grain size were truly uniform. Almost inevitably, however, the grains in a real metal are non-uniform and an additional parameter may be desirable to denote the spread of grain size; alternatively this could be given by regarding the grains to be a mixture of two or more uniform sizes. This case is often referred to as that of a mixed grain size. It should be borne in mind that the grains seen on a section through a specimen of uniform grain size will show a variation in area, according to whether the sectioning plane contains a maximum diameter D_M or is more or less close to the surface of the grain.

MEASUREMENT OF GRAIN SIZE

(i) *By Area:* The most used method of measuring grain size is based on comparisons of average areas of grains *as sectioned*. The number of grains appearing within some standard area at a fixed magnification is counted and from this, on the assumption of a uniform grain section, an equivalent area per grain is calculated. This area does not have any very direct physical significance, as should be clear from the discussion in the previous section. Quite often the connection with reality is strained even further, and this area is used to calculate the diameter of an equivalent circular area (or even of a square). This figure may then be called the grain diameter, but this cannot be regarded as a valid way of deriving grain diameter.

However, the concept of average area is useful and, indeed, widely used as a means of *comparing* grain sizes. It forms the basis of the A.S.T.M.[7] system of grain-size determination. The A.S.T.M. (micro) grain-size number g is defined by the equation

$$n = 2^{g-1} \tag{7}$$

where n is the number of grains in an area of one square inch of the specimen magnified to $\times$ 100. By rearranging equation (7), g can be expressed directly as

$$g = (\log n/\log 2) + 1 \tag{8}$$

If it is convenient to use some magnification M other than $\times$ 100, the value n_i counted at this magnification is converted to n as follows:

$$n = n_i\,(M/100)^2 \tag{9}$$

Counting grains in a standard area in order to determine g directly by the above equations (7)–(9) is known as the planimetric or Jeffries' method of determining the grain-size number. The area should have at least 50 grains in it and at least three fields should be counted. Parts of grains at the periphery of the enclosing frame are counted as half-grains. The procedure is simplified if a standard area of 5000 mm^2 is used (i.e. a circle of 79·8 mm dia.), in which case there is a factor f (or Jeffries' Number) for each magnification, which will convert the grain count to a figure giving grains/mm^2. Jeffries' Number is, of course, simply $M^2/5000$.

(ii) *A.S.T.M. Charts:* For characterizing grain sizes sufficiently well for many commercial operations, the A.S.T.M. grain-size number is estimated very simply by comparing the sample with a set of standard charts. These show an area (either schematically or using actual photographs of a standard), with grains chosen to be of a particular grain-size number. The charts may be in the form of either prints or transparencies; the prints are used with photographs of the specimens under test and the transparencies with images projected on the microscope screen. In either case, of course, the magnification used for the specimen should be $\times$ 100. If the specimen image is at some other magnification a correction $+q$ must be made to g using the formula:

$$q = 6{\cdot}64 \log (M/100) \tag{10}$$

There are four types of comparison standards, classified according to the appearance of the etch and the presence or absence of annealing twins in the structure. The aim is to use the type of standard which resembles the test specimen most closely.

The grain-size number is taken to be that of the standard which is closest to the test specimen; for most purposes, therefore, g is quoted as an integer (but for coarse grain sizes there are designations of zero or double-zero). Standards are available to half values of g, but it is often sufficient to state the grain size to be "between 6 and 7" or perhaps add "but nearer to seven". The authorized range of values of g in the A.S.T.M. standard is from 00 to 14; this is equivalent to ranging from an average intercept of 0·451 mm down to one of 0·00222 mm.

When the grain shape departs from being equiaxed it becomes difficult to assign a grain-size number by the comparison method and, indeed, the planimetric method itself is then of doubtful value. The A.T.S.M. standard suggests that an arbitrary procedure should be adopted which nevertheless provides some basis for comparing various samples. Three values of n are obtained for the longitudinal, transverse and through-the-thickness directions of the specimen as n_L, n_T and n_N respectively. The value of n used in equation (7) is then n_{AV} where

$$n_{AV} = 0{\cdot}7 \times n_L \times n_T \times n_N \tag{11}$$

The factor 0·7 is, in fact, only strictly appropriate to an equiaxed grain shape, and should tend towards unity as the grains become flatter. It must be concluded that this is not a satisfactory method of characterizing grains which are not equiaxed, except when making the most general comparisons.

To return to the comparison method, it is clear that it provides a quick way of assessing grain size with an accuracy which is sufficient in many commercial applications, but that it is not informative enough for interpretive experiments. It is, moreover, a subjective measure and gives scope for observer bias. It is even possible for a particular observer to prejudice his own estimate in favour of the number he chooses for the first comparison with the standard. This may be overcome to some extent by changing the magnification for each comparison of the three or more which should be made with each specimen, but this somewhat diminishes the advantage of speediness claimed for the method.

A similar series of standard charts is also published by A.S.T.M. for the measurement by comparison of *macro-grain* sizes. These numbers bear the prefix "M" and range from M0 to M15. The comparisons should be made at natural size (magnification × 1). A grain size of M0 is equivalent to an average lineal intercept of 32 mm and a grain size of 00 on the micro-grain scale is approximately equivalent to M12.

(iii) *Lineal Analysis:* A measure of grain size by lineal analysis is comparatively easy to obtain and may be used to represent various real attributes of an idealized grain structure. Moreover, if the correct procedure is used, a satisfactorily small standard error may result from readings which are speedily taken.

Lineal analysis for a single-phased alloy (which is what it amounts to for measuring grain size) is particularly simple, because all that is required is the total number of boundaries in a given length of line. This may be a real line drawn on a photograph or superimposed on an image from a reticle in a microscope ocular, or it may be an imaginary one defined by the amount of movement of a microscope stage.

If the grains are equiaxed, the orientation of the test line with respect to the specimen is not important and the measurement will be the same whatever direction is used. In general, there is some departure from the equiaxed shape in practice and to obtain a random sample the line should be applied over a series of directions 15° apart.*

There are two refinements to these precepts relating to the test line. The first is to use a series of parallel lines spaced more than one grain diameter apart (see Fig. 117). This is a particularly convenient arrangement to have on a reticle in a microscope ocular; the magnification is then chosen to give a suitable relative spacing of lines to the grain size. The second improvement which may be made is to use a circular test figure rather than a straight line. A circle automatically randomizes the direction and its use in an ocular turns out to be particularly good for minimizing eye strain. Hilliard[4] introduced this figure (p. 175) and has published details of its use, together with a nomograph for converting intercept density to grain size. It is found that eye strain is reduced and accuracy increased if the magnification is chosen to result in about 7–10 intersections of the circle with grain boundaries. If the number is much below 7, there is a tendency for some grains to be almost as large as the circle and, moreover, the counting requires a relatively large number of applications of the circle. On the other hand, if the number is much greater than 10, it becomes difficult to "tick off" the grains mentally as they are counted. The recommended diameter of the test circle is one no more than two-thirds of the field of view (otherwise it becomes confusing in recognizing grain boundaries at the edges of the field); the line thickness

* Although standards, such as the A.S.T.M. one, only specify two orthogonal traverses in any section.

of the circle should be no more than 0·01 mm. A reticle may be made by photographing, on a high resolution plate, a circle drawn with the correct proportions.

The circle is applied randomly to the test section and the number of intersections with grain boundaries N_1, N_2, N_n etc. counted. If a boundary appears to be tangential to the circle, a score of one is counted as for a clear intersection, but if it appears to pass through a triple junction the score is one and a half.

The results may be used either to compute the average lineal intercept $\bar{L}$ or an equivalent A.S.T.M. grain size number g_e. If the actual perimeter of the circle is P, its apparent or equivalent circumference on the specimen image will be reduced by the magnification M to P/M. Hence for one application of the circle

$$\bar{L} = P/MN \qquad (12)$$

To obtain $\bar{L}$ for several applications of the circle the total number of intersections $N = \sum_{n=1}^{n} Nn$ and the total line length nP must be used:

$$\bar{L} = nP/MN \qquad (13)$$

From this, it can be shown that a grain-size number which is approximately that which would be found using the A.S.T.M. charts is given by

$$g_e = -10{\cdot}00 + 6{\cdot}64 \log_{10} (nP/MN) \qquad (14)$$

The standard deviations, and thus the standard errors at a specified confidence level, are determined from the values of L_1, L_2, etc. for the n applications of the test circle. Alternatively, the standard error for g_e is determined. The standard error (or coefficient of variation) and the confidence level considered appropriate seem to be a matter of choice. Hilliard[4] in the data sheet issued by A.S.M. suggests that, in the absence of any specific instructions, grain sizes should be reported to an accuracy of $\pm$ 0·3 of a grain-size number; although he does not mention the confidence level explicitly, it appears to be the 68 per cent level, since his calculations are based on a standard error of $\sigma/N^{\frac{1}{2}}$. The equivalent coefficient of variation for the mean intercept $\bar{L}$ is $[100\ \sigma\ (L)]/\bar{L} = 10$ per cent. This accuracy demands surprisingly few readings; some 35 intercepts on an equiaxed, uniform, grain structure will suffice. Since grain structures so seldom appear to be equiaxed and uniform, it seems desirable to take rather more readings and, perhaps, to demand a higher confidence level. For experimental correlations, such as dependence of yield or of creep rate upon grain size, a confidence level of 95 per cent is suggested; this would usually require at least 100 intercepts to be measured.

(iv) *Area of Grain Boundary:* It is often of interest to have a measure of the amount of grain-boundary surface in a specimen. This is very simply found, because S_v, the grain-boundary area per unit volume is given by

$$S_v = 2N/L = 2/\bar{L} \qquad (15)$$

that is, S_v is equal to twice the number of intersections per unit length of a lineal traverse. This, of course, is only true for a single direction of traverse when the structure is equiaxed. However, if N is based on a random traverse (such as that obtained with a circular test figure), the result holds for structures which are not equiaxed. In other words, equation (15) is true for random traverses of any space-filling arrangement of shapes.

The truth of equation (15) is not at all obvious intuitively and the proof rather too long to include here. The interested reader will find it in a paper by Smith and Guttman.[8]

(v) *Grains in Unit Volume:* The number of grains in unit volume is related to S_v by an equation of the form [9,10]

$$S_v = FN_v^{\frac{1}{3}} \tag{16}$$

where F is a constant which depends upon the shape of the grain. It has been suggested that the value for real grains—approximately tetrakaidecahedra—is 8/3.

RELATIONS OF L TO OTHER PARAMETERS ("DIAMETERS")

In dealing with the determination of grain size by the measurement of lineal intercept, the results have been left in the form of the average lineal intercept $\bar{L}$, and this is, indeed, a useful parameter for comparison of grain sizes. Also, its direct relation to S_v (the area of grain boundary per unit volume) makes $\bar{L}$ of added utility. However, as indicated in the introductory remarks to this section, particular problems may demand that a particular aspect of the dimensions of an average grain is important.

"Average grain diameter" is often quoted, but unless this is linked to a stated grain shape the measurements do not have any clear meaning. For example, the following equations define an average diameter $\bar{D}_c$ for cubic grains, $\bar{D}_t$ for tetrakaidecahedral grains, and $\bar{D}_s$ for spherical grains

$$\bar{D}_c = 2{\cdot}25\ \bar{L} \tag{17}$$

$$\bar{D}_t = 1{\cdot}776\ \bar{L} \tag{18}$$

$$\bar{D}_s = 1{\cdot}6\ \bar{L} \tag{19}$$

Equations (17) and (18) are as quoted by Hilliard[11]; Hensler[12] in a recent experimental and theoretical study concludes that $\bar{D}_s = 1{\cdot}6\ \bar{L}$ whereas Hilliard quotes $1{\cdot}5\ \bar{L}$.

When the distance between grains through a grain-boundary face is important, the diameter of a face is the relevant parameter. Again, the scaling factor depends on the geometry. For a tetrakaidecahedron the diameter $\bar{D}_F$ of one of the hexagonal faces is given by

$$\bar{D}_F = 1{\cdot}185\ \bar{L} \tag{20}$$

This distance is also twice the length of a grain edge (i.e. the line of a triple junction). Average grain-diameter might also be interpreted to refer to the average diameter $\bar{D}_A$ exposed on the polished section; this is related to the average area and therefore to the A.S.T.M. grain-size number. It has been shown that for spherical grains

$$\bar{D}_A = 1{\cdot}27\ \bar{L} \tag{21}$$

MIXED GRAIN SIZES

Since there are only very limited grain shapes which can fill space and also obey the other restrictions imposed by surface tension, it is not likely that a piece of real metal will contain uniform grains of a particular shape. It seems much more likely that a large proportion of grains might approximate to this ideal and the inevitable interstices be filled with less regular grains of other sizes.

It is reasonably simple to examine the effect of such mixtures on the appearance of sections through aggregates of spheres. It can be shown that the ratio of A_{max}, the maximum area revealed in a plane section, to $\bar{A}$, the average area, is 1·5 if the spherical grains are all identical. For grains of a more realistic shape the value of A_{max}/A lies between 1·7 and 2. Thus a value of this ratio which is greater than, say, 4 indicates a marked departure from uniformity of grain size; in fact, it would be obvious that this is so by visual comparison of such a sample with a standard equiaxed specimen.

When the mixture is sufficiently heterogeneous the principal grain sizes will also become obvious as separate peaks on the frequency histogram of intercept or areal value. Such bimodal distributions (it is very rare for there to be more than two peaks) indicate the operation of some special factor such as temperature or composition, or inhomogeneity of deformation.

TWIN AND SUB-GRAIN BOUNDARIES

For most purposes a meaningful grain size excludes twin or sub-grain boundaries from the counting. It is generally easy to ignore sub-grain boundaries because, in fact, they require special techniques to make them show up clearly. Annealing-twin boundaries, however, may easily be confused with normal boundaries and thus lead to an apparently smaller grain size. This is particularly the case for twin boundaries which completely cross grains. The use of polarized light and a sensitive tint plate to render the grain contrast in colour is very helpful in distinguishing twin boundaries from normal ones. It has been shown in the author's laboratory that independent observers agree very closely on the true lineal intercept using coloured contrast but either disagree to a significant extent or slow down considerably with the same specimens etched for viewing in normal green light. Of course, the particular metal or alloy containing twins must either be sufficiently anisotropic or be capable of being rendered sensitive to polarized light. Fortunately, this can be done for two of the principal groups of alloys which show twins—those of copper and lead.

SUMMARY

It will be seen from the above discussion that there is no single parameter which can be used universally to characterize "grain size".

For a great many practical, everyday, applications the A.S.T.M. grain-size number, which is based on a planimetric (areal) formula, will suffice. The A.S.T.M. number may be estimated, very simply, by comparison with standard charts.

When a more precise measurement is required, particularly one to which a statistically determined error is attached, then lineal analysis should be used. To allow for the presence of a non-equiaxed structure, the direction of traverse should be made random, preferably by using a circle as a test figure. Under normal conditions a 10 per cent accuracy is then obtained with comparatively few intercepts counted (35 to 120 according to the confidence level chosen). This method is therefore rapid, and accurate to a degree which can readily be computed and which should be stated.

The average lineal intercept is simply related to various other parameters which may be of interest, namely, the grain-boundary area per unit volume, the average grain volume, the average "caliper" diameter of the grain and the average diameter of a grain face. These relations are specific to an assumed grain shape; probably the best approximation for these purposes is the tetrakaidecahedron.

Other Quantitative Observations

There is no point in trying to cover all the possible kinds of measurement that could be made in metallography. Many of these are unique to particular problems and, once done, do not need to be done again. Others are very well known and described elsewhere.

A large group of these is concerned with orientation determination, either of crystals or of their boundary planes. Slip trace analysis might be considered in this group. These matters are excellently treated in Barrett and Massalski.[13]

Angular relationships are also often required—angles between grain-boundaries and tensile directions, dihedral angles at grain-boundary junctions or between various phases, etc. If large numbers of readings are necessary to obtain frequency histograms, such measurements are best done on the microscope using either a graduated rotating stage and crosswires in the ocular or, possibly, a special ocular which has a movable protractor built into it. For the measurement of entities of varying length, for example, cracks, the ocular from a diamond pyramid or Knoop hardness machine is quite useful. The movable shutter in this is used to gauge the crack length, which is read off on the vernier scale in the usual way.

A somewhat cruder method of measuring some features is based on the plates used for sizing knitting needles. A series of holes or slots (according to the general shape of the feature) is made in a stout card, one hole for each class in the expected range for the frequency histogram. The template is moved over the screen of the microscope on which the specimen is imaged, and the class number of each feature rapidly assessed on a "go—no-go" basis.

Another technique of considerable use is to project the image of the specimen, by means of an auxiliary prism and an external mirror, on to a sheet of paper held in a fixed position on the bench beside the microscope. The feature of interest—say, a marker line across sliding boundaries—is drawn and the measurements made at leisure with a ruler, protractor or set-square, as appropriate. At first sight this might not appear to offer any advantage other than some saving of time (which is always pleasant to achieve). However, particularly on deformed specimens where the surface is rumpled, it does permit continual refocusing, or even changes in lighting to be made, to clarify obscure features. Unless a photograph is taken of each feature, rather than of entire regions of the specimen, this accuracy of focus cannot be achieved by photography. Since several hundred observations are usually required, the tracing method has a distinct advantage in accuracy as well as in time and cost. Another convenience which both photographs and tracings have is that they provide a permanent record which can be rechecked; they may also be readily used for independent observers to measure an identical set of features. In this connection, the use of a simple replica (see Chapter 9) is worth noting.

Appendix

Simple Statistics

In quantitative metallography, the measurements made are seldom of much use until they have been manipulated in some way, to extract parameters which can then be taken to characterize the feature examined. This is because there is generally scatter due to the methods of sampling or of measuring and often scatter due to an inherent variability of the features studied. The readings therefore cover a more or less wide range of values and it is the aim of mathematical statistics to provide the means of making statements about the results, or of comparing one set of results with another, with a precision which is defined. Since the particular circumstances relating to the readings may dictate what kind of statement—or *statistic*—is used, this section sets out to summarize some elementary ideas about statistics. As so often has happened in this book, it is not possible to be both brief and rigorous. These notes should be used, therefore, with caution and perhaps applied more to the interpretation of results using established procedures than to formulation of statistical treatments for new kinds of measurements. There are many textbooks designed to assist the scientist and engineer in applying statistics to his problems; some of these are listed in refs. 14–16. It must be emphasized that the value of statistical methods is to assist the planning of meaningful experiments, not to extract information from chaos.

Mean Values

The collection of readings obtained for a particular parameter, such as grain diameter, particle diameter or crack length, is known as *the sample* (e.g. the diameters of those inclusions revealed on one particular area of one section of a specimen). It is considered to be taken from the entire range of possible readings, called the population (e.g. the diameters of all the inclusions in the whole of the specimen from which the sample was taken). A usual and useful way of characterizing a collection of measurements is by their mean values, but it should be appreciated that there are several kinds of mean. The simplest and most widely used is the *arithmetic mean*, obtained by summation of all the readings divided by the total number of readings. If x_m is a typical measurement, then the arithmetical mean of n such measurements is given by

$$\bar{x} = \left(\sum_{m=0}^{n} x_m\right) \Big/ n \qquad (22)$$

Some writers define the mean in terms of a convenient method of computation and graphical representation, based on the concept of grouping results in *classes*. To appreciate this, it is necessary to return to a consideration of the sample as a whole. It is often useful to display the results in a manner which shows how often certain values occur. This is a *frequency histogram*. Fig. 122 (*a*)

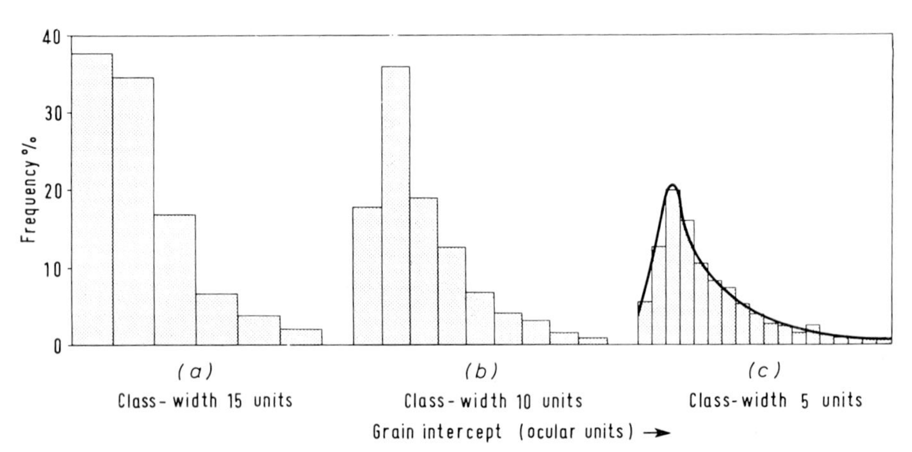

Fig. 122. (*a*) Histogram of frequencies of lineal intercepts L of grains on a section of lead.

(*b*) and (*c*) show the effect of successively diminishing the class width of L, so that the histogram is approximately equivalent to the enveloping curve.

shows a histogram of measurements of lineal intercepts of grains on an annealed specimen of lead. The horizontal axis shows values of the intercept (in units read on an ocular graticule) grouped into classes of width 10 units (i.e. 1–10, 11–20, etc.). The vertical axis represents the frequency with which readings fell within the various classes. To convey the idea of classes, the diagram uses rectangles to represent frequency distribution between classes. The class having the highest frequency is spoken of as having the *modal value*; it is the most probable value of grain intercept in this population. In metallography the modal value is not generally very significant.

The mean defined as in equation (22) is not directly calculable from data presented in the manner shown in Fig. 122 (*a*). Instead, a good approximation sometimes called the weighted mean $\bar{x}_w$, can be derived by adding up all the products of frequency times mid-value for the classes and then dividing by the total of frequencies. In symbols, if the classes have frequencies f_1, f_2, f_3 . . . f_m and their mid-values are x_1, x_2, . . . x_m, the weighted mean is

$$\bar{x}_w = \left[\sum_{m=1}^{n} (f_m x_m)\right] \Big/ \sum_{m=1}^{n} f_m$$

or

$$\bar{x}_w = \left[\sum_{m=1}^{n} (f_m x_m)\right] \Big/ n \qquad (23)$$

When the width of the classes is made smaller and smaller, $\bar{x}_w$ approaches $\bar{x}$, becoming equal to it when the class width is one unit of measure for x. At the same time, the histogram in Fig. 122 (*a*) becomes more closely approximated by an enveloping curve—as shown in Figs. 122 (*b*) and (*c*). The useful class width which may be employed in a particular case depends upon the way the readings are obtained and upon their number. For example, if readings of an angle are being made and the protractor only reads to the nearest 5°, it is clearly meaningless to have classes of width smaller than 5°. On the other hand, even if the protractor reads to 1°, it might equally be destroying the usefulness of the histogram to have a class width of 1° for a total number of readings less than 90.

The Normal Distribution

In many cases (but not in Fig. 122) the curve enveloping the histogram is a special one which is basic to the statistics we shall consider. It is the curve of *normal* or *Gaussian* distribution of values—Fig. 123. This kind of distribution is very common and reasons for this may be surmised from an inspection of it. Thus, the modal value (the most probable value) is equal to the mean, because of the symmetry of the curve; again, values close to the mean are also highly probable (their frequency is large). As values successively removed from the mean are considered, their frequencies diminish increasingly rapidly; that is, their probabilities decrease. However, it is not impossible—though highly improbable—to obtain very small or very large readings out on the "tails" of the curve.

For our purpose it is the shape of the Gaussian distribution which is its unique property; as shown in Fig. 123, we can recognize it with differing means and overall widths (or *dispersions*). Clearly, too, a description of distributions related to the means and dispersions of equivalent Gaussian distributions would seem to be an apt manner of characterizing and comparing them. We have already spoken of the mean and indicated that the arithmetic mean is a preferred statistic; we shall now consider measures of dispersion.

Measurement of Dispersion

The dispersion may be measured in a number of ways. The simplest would be to indicate the scatter by stating that all the readings fall in the *range* $x_{\min}$ to $x_{\max}$. However, this is not as informative as may often be desirable, for it fails to tell us anything about the probability of occurrence of various readings; it may also be misleading because it puts greater reliance upon those values most infrequently found (in the tails of the distribution).

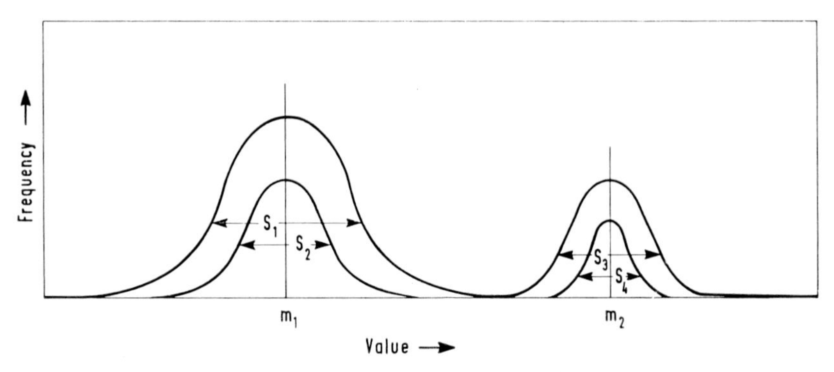

Fig. 123. The curve of the normal or Gaussian distribution, showing how the characteristic shape is preserved with curves of differing means (m_1 and m_2) or dispersions (here characterized by s_1, s_2, etc).

A more useful statistic is the *mean deviation* which is defined as the mean of the numerical values of deviations from the mean. The deviation of a value x_m from the mean is $d_m = x_m - \bar{x}$. The mean deviation, because it considers numerical values of d_1, d_2, etc., has a positive value

$$\bar{d} = [|d_1| + |d_2| + \ldots + |d_m|]/n \tag{24}$$

The algebraic sum of the deviations for a Gaussian distribution is, of course, zero. This provides a simple way of obtaining the mean with a reduced amount of arithmetical work. By inspection of the data a mean is guessed. Then the difference Δm between each value x_m and m is tabulated. Thus

$$x_m = m + \Delta m$$

so

$$\sum_{m=1}^{n} x_m = nm + \sum_{m=1}^{n} \Delta m \tag{25}$$

or

$$\bar{x} = (\sum \Delta m)/n$$

In this way, the mean $\bar{x}$ may be found by adding the mean of the differences Δm to the assumed mean m, remembering to sum Δm algebraically.

However, the most widely used measure of dispersion is the *standard deviation*, which is defined in terms of the squares of the deviations. The standard deviation of a population is usually denoted by σ, and thus

$$\sigma^2 = (d_1^2 + d_2^2 + \ldots + d_m^2)/n \tag{26}$$

The square of the standard deviation, σ^2, is important enough to have a name of its own—the *variance*. It will be seen that the standard deviation, by its dependence upon the squares of deviations, gives emphasis to the degree to which values bunch near to the mean or to which they lie out in the tails. Clearly, a small value of the standard deviation indicates a comparatively narrow distribution.

It is important to be clear that, when dealing with a sample, the calculation of a standard deviation from the results is a process of *estimating* the standard deviation of the population. From a limited number of observations—the sample—the calculation is attempting to derive the likely standard deviation for the whole population from which the sample is drawn. In view of this, it is necessary to make a slight alteration to equation (26), replacing the divisor n by $(n - 1)$. To emphasize that it is an estimate of the standard deviation, we write S instead of σ. Thus equation (26) becomes

$$S^2 = (d_1^2 + d_2^2 + \ldots + d_m^2)/(n - 1) \qquad (27)$$

Two reasons for changing to $(n - 1)$ can be given. Firstly, it will be seen that the value of S will always be less than σ unless the sample and population means are identical; thus, in the absence of a knowledge of the population mean, using $(n - 1)$ compensates for this underestimation. Secondly, for a sample containing only one observation, equation (26) would give a standard deviation of zero; use of $(n - 1)$ renders this extreme case indeterminate, which is intuitively to be expected. When n is large (say $n > 100$), the difference between using n and $(n - 1)$ becomes relatively unimportant. The estimate of the standard deviation is important in comparing sets of data; some of these uses will be described in later sections.

The determination of an estimate of the standard deviation often involves a large amount of tedious arithmetic. There are two formulae which may be used to shorten this. The first again makes use of the idea of an assumed or working mean, m. From this the sum of the squares of the deviations is calculated as

$$\sum_{m=1}^{n} (x_m - m)^2$$

and the mean of this, denoted by D^2, is found as

$$\left[\sum_{m=1}^{n} (x_m - m)^2\right] \Big/ (n - 1).$$

It can be shown that the variance is then given by

$$S^2 = D^2 - (m - \bar{x})^2 \qquad (28)$$

The other formula is in a form which is particularly suitable for computations on certain desk calculators. It is

$$S^2 = \left[\sum_{m=1}^{n} (x_m)^2 - (\sum_{m=1}^{n} x_m)^2/n\right] \Big/ (n - 1) \qquad (29)$$

It is possible to set up Σx_m and $\Sigma (x_m)^2$ in one operation on some calculators; the formula just quoted then becomes very simple to use in computing S^2.

Sheppard's correction should be applied to the estimate of the standard deviation when the data have been grouped into classes. It is in a form which is readily applicable to the derivation of S by the use of the assumed mean. If the class width is c, then D^2 should be reduced by subtracting $c^2/12$. If the second formula is used $m = 0$, and thus $D^2 = \bar{x}^2 = (x_m)^2$, the correction of $c^2/12$ then has to be made to $(x_m)^2$ before dividing by n. In a great many cases the value of $\bar{x}^2$ is very much greater than $c^2/12$, especially when c itself is small, and the neglect of the correction does not introduce a serious error.

USE OF STANDARD DEVIATION

The standard deviation has a special significance in relation to the normal distribution, which has not been brought out in the foregoing discussion. The equation for the curve of the normal distribution contains σ^2 in such a way that the probability of an observation falling between $\bar{x} \pm \sigma$ is 0·68 or between

$\bar{x} \pm 2\sigma$ is 0·95. These are, in fact, the areas under the parts of the curve between $\pm\sigma$ or $\pm 2\sigma$ (see Fig. 123). Thus, these zones are sometimes called the 68 and 95 per cent confidence limits respectively, implying that there are 68 or 95 chances per 100 that a result lies in this zone.

This leads to the idea of a *standard error of the mean*, quoted to some (percentage) confidence range. It is clear that if various random samples are taken from a normal population (which is infinite) each sample will have a mean that will not, in general, be that of the population. As the number n in the sample becomes larger, the difference between the sample mean and that of the population would be expected to become smaller. In fact, it can be shown that the *distribution* of these means for a given value of n is itself a normal one having a standard deviation $S/n^{\frac{1}{2}}$ (S still being the estimated standard deviation of the population as a whole). We may therefore write that a mean $\bar{x}$ determined from a particular sample has a standard error of $\pm S/n^{\frac{1}{2}}$. Following the ideas expressed in the preceding paragraph, we can say that the probability that the mean lies between the limits given by $\bar{x} \pm S/n^{\frac{1}{2}}$ is 0·68; similarly for $\bar{x} \pm 2S/n^{\frac{1}{2}}$, the probability is 0·95. In general, probabilities may be assigned to any limit of the form $tS/n^{\frac{1}{2}}$, through the use of a table of values for the parameter t. This has special uses in statistical work other than the one indicated here. The table is usually called "Percentage points of the t-distribution" and gives values of t for combinations of ν, the number of degrees of freedom and P, the probability. Some selected values of t are given in Table 6. The number of

TABLE 6

Selected Values of t

Degrees of Freedom	Probability P		
	10	5	0·2
5	2·02	2·57	5·89
10	1·81	2·23	4·14
20	1·72	2·09	3·55
40	1·68	2·02	3·31
120	1·64	1·96	3·09

degrees of freedom is not a simple concept, and involves considerations of the type which were quoted to justify using $(n - 1)$ instead of n in the estimate of the standard deviation; in the kind of work involved in quantitative metallography ν is almost always equal to $(n - 1)$. When considered in the context of determination of standard errors, P is the probability that the mean is *outside* the zone defined by the limits quoted. In other words, the confidence level is the percentage giving the probability $(100 - P)$ that the mean is within the zone. For example, if P in the table of t-distribution is 5, then the confidence level is 95 per cent. A 95 per cent confidence level is usually regarded as satisfactory in metallurgical work, but, naturally, circumstances may sometimes dictate a more stringent level or sometimes a more relaxed one.

On a diagram of values of $\bar{x}$ against some other parameter, the standard error is conveniently plotted as an *error bar* by giving the point for $\bar{x}$ extension (above and below $\bar{x}$ in the x-direction) equal to the standard error; the diagram should, of course, bear an annotation which calls attention to the confidence level and to the size of the sample.

An alternative method of defining the error of a mean is to quote the standard error of the standard deviation itself. This is $S/(2n)^{\frac{1}{2}}$ and therefore the result would be written as

$$\bar{x} + [S\{1 \pm 1/(2n)^{\frac{1}{2}}\}]$$

Other Frequency Distributions

Although the normal distribution is a very usual one, it does happen that data are sometimes better fitted by some other distribution. It may merely be that a transformation of the data in some simple manner will convert them to a normal distribution; for example the logarithms of the data might form a normal distribution. This is spoken of as a *log-normal distribution*.

When the distribution is markedly asymmetrical it is called *skew*. Other distributions are the *binomial* and the *Poisson*. The binomial distribution is a very general one, of which others are special cases. The Poisson distribution is the special case of the binomial which relates to events which have a low probability (and therefore happen rarely) but are observed with appreciable frequency because the number of trials is very large. The binomial distribution with higher probabilities of occurrence and with a large number of trials (not necessarily very large) gives a histogram which is approximately enveloped by the normal distribution curve. It should be noted that the *normal distribution of the means* (which formed the basis of determining the standard errors of the means) is found even for sample distributions which themselves depart appreciably from the normal. This would appear to be the basis of the assumption implicit in much work on quantitative metallography; namely, that the means may indeed be treated as if they are based upon normal distributions.

Some care is necessary when comparing a log-normal distribution with the same figures treated as a (skew) normal one. For example, the mean of the logarithmic values is clearly the mode of the "normal" values, not their mean. Similarly, one cannot take the logarithmic mean $(\overline{\log x})$ and logarithmic standard deviation $(S_{\log x})$ and derive a standard error of the arithmetic mean $\bar{x}$ as $\bar{x} \pm S_{\log x}/n^{\frac{1}{2}}$. If such comparisons are desired the following conversions may be used:

$$\log \bar{x} = \overline{\log x} + 1 \cdot 1513\, S_{\log x} \text{ for the means}$$

and $S_{\log x} \simeq 0 \cdot 4343\, S_x/\bar{x}$ for the standard deviations. (The numerical factors here arise from $\log e = 0 \cdot 4343$.)

Significant Differences

Although this subject is a large one and requires a more advanced treatment than is possible in these notes, it is an important one; moreover, the rules for a mechanical operation of the tests are not themselves difficult to follow. The full analysis to test significance may be tedious, and there are various so-called non-parametric tests which are useful as preliminary probes.

One of these which is suited to the kind of results found in quantitative metallography is illustrated in Table 7. The measurements are supposedly of the diameters of a precipitate observed on a polished section; both specimens have had identical heat treatments, but specimen B contains a trace addition thought to affect ageing and therefore the size of the precipitate. The observations are placed in two columns side by side, as shown in Table 7. Against every observation of B a sign is placed, a plus sign if it is greater than its neighbour in A, a minus if it is smaller and a zero if it is equal. The difference $|P - M|$ between the sum of the plus signs and the sum of the minus signs is found and compared with twice the square root of their total, $2\sqrt{(P + M)}$. (The zeros are ignored.) Then, if $|P - M| > 2\sqrt{(P + M)}$ it can be taken as unlikely that the two samples of observation have been drawn from the same population. In other words, it is likely that the trace addition has had a significant effect. In Table 7 $|P - M| = 12$ and $2\sqrt{(P + M)} = 8{\cdot}5$, so there

TABLE 7

Diameters of Precipitate Particles in Two Samples

A	B
1·1	1·3 +
0·8	1·0 +
1·2	1·3 +
1·2	1·4 +
1·1	1·4 +
0·9	1·2 +
1·5	1·2 —
1·3	1·5 +
1·0	1·2 +
1·4	1·5 +
1·6	0·9 —
1·5	1·6 +
1·9	2·0 +
1·0	1·0 O
1·0	1·0 O
0·9	1·0 +
1·4	1·5 +
1·1	1·3 +
1·0	1·2 +
1·3	0·8 —

seems to be a case for considering the hypothesis about the effect of the trace element as likely. It is worth while, therefore, making the full analysis.

The way in which this is done rests upon the notion of testing the hypothesis that both samples are drawn from the same population. To do this, the variances S_A^2 and S_B^2 for each of the samples are estimated and the expression

$$Q = |\bar{x}_A - \bar{x}_B|/(S_A^2/n_A + S_B^2/n_B)^{\frac{1}{2}} \tag{30}$$

is computed; $\bar{x}_A$ and $\bar{x}_B$ are the means of samples A and B respectively and n_A and n_B their numbers of observations. Q is then compared with the value of t appropriate to the smaller of the two numbers of observations and with the confidence level chosen. For example, if $Q > t_{95}$, then the two samples have means significantly different at the 95 per cent confidence level (in the case when n_A and n_B are both greater than 120, $t_{95} = 1{\cdot}96$).

Returning now to Table 7, we find that $\bar{x}_A = 24{\cdot}2$ and $\bar{x}_B = 25{\cdot}3$; $S_A^2 = 26{\cdot}8/19$ and $S_B^2 = 29.3/19$; thus $Q = 2{\cdot}8$. For $n_A = n_B = 20$, the value of t at the 98 per cent confidence level is $2{\cdot}53$. Hence we can deduce that there is only one chance in 50 that the samples are from the same population. However, if we set the confidence level at 99·8 per cent, $t = 3{\cdot}55$ and Q is less than this: it cannot now be concluded that there is a significant difference in the two samples at this level of confidence (which is 1/500 that the samples differ by the observed amount through chance alone).

Summary

1. The most useful measures of a sample of observations drawn in a random manner from all possible such observations (the *population*) on a specimen are the mean $\bar{x}$ of the data and some measure of its dispersion.

2. If it is assumed that the population is a normal distribution, the *standard deviation* S of the population, estimated from the sample, is the most useful measure of dispersion. It is given by

$$S^2 = (x_m - \bar{x})^2/(n - 1)$$

There are variants of this formula which are more convenient for computation in certain cases.

3. The mean can be quoted to lie within limits defined by a standard error with a specified confidence. Thus it can be written as

$$\bar{x} \pm tS/n^{\frac{1}{2}}$$

where t is a parameter chosen in conformity with the desired *confidence range* and the value of n.

4. Samples may be compared in several ways to see whether they are drawn from the same population. A simple method of comparison at an assigned confidence level is comparison of the standard error of the difference between the means with a chosen value of t.

5. It should always be borne in mind that the above statements may only be strictly true for a normal distribution. It may be necessary to process the raw data in some way to force it into the form of a normal distribution—e.g. by taking the logarithms of the observations.

6. *Warning.* This summary and the notes on which it is based are meant to be guides to the interpretation of published results and to the processing of data closely similar in nature to these. *If in any doubt, consult a text on statistics.*

REFERENCES

1 C. S. Smith, *Metal Interfaces*, p. 65, 1952, Cleveland (Amer. Soc. Metals).
2 C. S. Smith, *Metallurgical Reviews*, 1964, **9** (33), 1.
3 R. T. DeHoff and F. N. Rhines, *Quantitative Microscopy*, 1968, N.Y. (McGraw-Hill).
3*a* J. E. Hilliard and J. W. Cahn, *Trans. A.I.M.E.*, 1961, **221**, 344.
4 J. E. Hilliard, *Metal Progress*, 1964, **85,** 99.
5 R. L. Fullman, *Trans. A.I.M.E.*, 1953, **197,** 447 and 1267.
6 A.S.T.M. Standards, 1966, Part 31 (No. E 45–63), p. 126.
7 A.S.T.M. Standards, 1966, Part 31 (No. E 112–61, revised 1963), p. 222.
8 C. S. Smith and L. Guttman, *Trans. A.I.M.E.*, 1953, **197,** 81.
9 H. F. Kaiser, *Metals and Alloys*, 1938, **9,** 23.
10 E. E. Underwood, *Metals Eng. Quarterly*, 1962.
11 J. E. Hilliard, G.E. Report 62-RL-3133M., Dec. 1963 (Schenectady, New York).
12 J. H. Hensler, *J. Inst. Metals*, 1968, **96,** 190.
13 C. S. Barrett and E. Massalski, *The Structure of Metals*, New York, 1966 (McGraw-Hill).
14 J. Topping, *Errors of Observation and Their Treatment*, London, 1961 (Chapman and Hall).
15 R. Loveday, *Statistics, a First Course* and *Statistics, a Second Course*, London, 1961 (Cambridge University Press).
16 J. Moroney, *Facts from Figures*, London, 1951 (Penguin Books).

11 A Compendium of Complementary Techniques

There are a great many techniques which are part of metallography but not appropriate to describe in detail in a text on the optical microscope and its variants. These are listed here with a short description of the kind of information they yield.

1. Electron Microscope

Three standard types of electron microscope are generally available commercially today. Small microscopes working at ~ 50 kV are relatively inexpensive (not greatly different in price from the most expensive optical microscopes). They are also much less complex and much easier to use than the larger models. Their scope is limited to examination of replicas at magnifications up to about × 50,000, when the resolution would be of the order of 50 Å. The most used type of microscope in metallurgy works at a maximum of 100 (or perhaps 150) kV and has a resolution of better than 10 Å at × 120,000. These microscopes cost three or four times as much as the first group and require expensive ancillary equipment such as air conditioning, water filtration, etc. They are also used for replica work, but a major technique in metallurgy is the examination of thin foils of metals. These foils are no more than ~ 1000 Å thick and are then sufficiently transparent to electrons that defects (dislocations) and precipitates may be clearly resolved. The information obtained by imaging using electron-optical techniques is greatly extended by combining it with electron diffraction of the same thin foils. In this way crystallographic information about orientation relationships, phases and dislocation is linked to the more conventional-looking image.

The resolution and versatility of the electron microscope, already great, is being further developed at the time of writing by increasing the accelerating voltages to 500 and 1000 kV. Such instruments are exceedingly expensive, and bear the same relation to the optical bench microscope that the Mt. Palomar telescope does to a good pair of binoculars.

2. Reflection Electron Microscope

Some electron microscopes can be adapted to view specimens with oblique incidence, thus forming a kind of "taper section" shadowgraph of the surface examined. These conditions lead to a resolution of a few hundred Ångstroms. Useful information which is beyond the limit of the optical microscope may thus be obtained.

3. Scanning Electron Microscope

Where high resolution and high magnification are required combined with a large depth of field, the scanning electron microscope is the instrument to be used. An example where it can be invaluable is in the examination of deeply indented fractures, particularly those which, by their surface fibrosity, make replication difficult.

The instrument scans the surface with a focused beam of electrons and the image is formed by collecting the back-scattered electrons. These are converted to an optical image through a scintillation counter, amplifier and cathode-ray tube. Resolution is once again a few hundred Ångstroms.

4. Electron-probe Analyser

Another use for a focused beam of electrons is to make a chemical analysis of very small regions of micro-specimens, employing the X-rays excited by the incident electrons. The diameter of the probing beam is of the order of a micrometre or less, so that very small areas may be analysed. The area analysed may be identified by also using the electron beam to scan the specimen and form an electron or X-ray image. These instruments are rather more expensive than the general run of medium-power electron microscopes (100 kV) and, like most of these complex electronic instruments, require trained staff to operate and service them.

5. Electron-emission Microscope

The conventional electron microscope referred to earlier in this chapter uses a beam of electrons focused on to the specimen and is analogous to the optical microscope. It is also possible to obtain a useful image from electrons emitted from a specimen. These electrons are drawn off from the specimen by a suitably high applied voltage, but at lower temperatures some further assistance is required; either the specimen is coated with a substance such as barium, or it is bombarded by ions. Its most useful application is therefore in the examination of changes which occur at elevated temperatures.

6. Field-emission Microscopes

Yet another way in which a useful image can be obtained with electrons is in the field-emission microscope. The specimen is limited to having the form of a fine wire polished to a point of radius $\sim 1\ \mu m$. A high voltage (10 kV) is employed to draw electrons from the tip of the wire and accelerate them, within an evacuated chamber, to impinge on a fluorescent screen concentrically placed with respect to the wire tip. The purely geometrical magnification obtained is about × 100,000, and the resolution is better than 50 Å. The information gained is entirely limited to the surface of the wire and distinguishes the presence of deliberate or naturally occurring layers of oxides, absorbed materials, etc.

7. X-ray Microscopy

The familiar application of X-rays in radiography may be extended to the formation of images of micro-specimens—microradiography. However, because no lenses have been developed for imaging X-rays, the magnifications achieved are no more than a few times. The technique permits some identification of chemical species, and may be used to identify inclusions and void-like defects. Its principal advantage over optical microscopy is perhaps that the radiograph records information from the whole of the radiated volume, which could be a slab of material of the order of 0·1 mm thick.

X-rays may also be diffracted by micro-specimens (as distinct from being absorbed, as in radiography) to form images which give information about crystal defects in the specimen. The technique depends upon variations of crystal structure and their effect upon the diffraction of X-rays. A modification of the technique is to use a glancing angle of incidence to form an X-ray topograph of the surface. Magnifications are again small, being limited to those obtained by purely geometrical arrangements involving a divergent beam. Moreover, the technique is also mainly limited to pure metals.

A third metallographic application of X-rays is that of fluorescence analysis. A primary beam of X-rays is used to excite characteristic fluorescent X-radiation, which is analysed using a spectrometer. High accuracy may be realized in the analysis of relatively small samples, although the size of the area analysed is more than an order of magnitude larger than can be examined with the electron-probe analyser.

8. Field-ion Microscopy

The most spectacular resolution of any microscope is that obtained with the field-ion microscope, namely about 3 Å. This is sufficient to be able to "see" individual atoms, and the magnification is a million or more. As in the field-emission microscope, the specimen must be in the form of a fine wire with a polished tip. Atoms of an inert gas are ionized by reactions with atoms of the metal tip and then accelerated to cause scintillations on a fluorescent screen. Crystal defects, grain boundaries and some alloying and segregation phenomena are capable of being studied using this technique. As in all of these techniques—and in optical microscopy—there are clear limitations to what can be done using them; the limitations in field-ion microscopy are particularly severe.

9. Other Metallographic Techniques

The complementary techniques listed represent a family, most of which form some kind of image of the specimen. Moreover, interpretation of this information often turns upon referring it back to the more familiar optical image. There are many other tools the metallographer has at his disposal, some of these perhaps being more aptly brought under the heading of non-destructive testing. In order to complete the picture these are mentioned; this may also assist in putting the whole field in perspective.

Hardness testing is an invaluable metallographic procedure, and its extension to micro-hardness testing perhaps brings it into the ambit of optical metallography. Crack detection using magnetic or fluorescent fluids, ultrasonic reflections or eddy currents could also be regarded as a combination of a special kind of etching with macro-metallography.

Thus we close this book as we commenced it by calling attention to the range of techniques available to the metallographer, but the intervening pages should have amplified this statement and emphasized the limitations and advantages of the various methods.

Select Bibliography

OPTICAL METALLOGRAPHY

1 *Metallography in Colour*, Philadelphia, 1948 (Amer. Soc. Test. Mat.).
2 G. L. KEHL, *Principles of Metallographic Laboratory Practice* (3rd edn), New York, 1949 (McGraw-Hill).
3 D. McLEAN, *Metal Treatment*, 1951, **18,** 51.
4 E. C. W. PERRYMAN, *Metal Ind.*, 1951, **78,** 22, 51, 71 and 111.
5 *Refresher Course: The Microscopy of Metals*, London, 1953 (The Institution of Metallurgists).
6 S. TOLANSKY in *Properties of Metallic Surfaces*, London, 1953 (Inst. of Metals).
7 *Light Microscopy*, Philadelphia, 1953 (Amer. Soc. Test. Mat.).
8 *Modern Research Techniques in Physical Metallurgy*, Cleveland, 1953 (Amer. Soc. Metals).
9 R. H. GREAVES and H. WRIGHTON, *Practical Microscopical Metallography* (4th edn), London, 1956 (Chapman and Hall).
10 R. C. GIFKINS, *J. Aust. Inst. Metals*, 1958, **3,** 143.
11 B. CHALMERS and A. S. QUARREL, *Physical Examination of Metals: Vol. 1 Optical Methods*, London, 1960 (Arnold).
12 *Metallography—1963 (Proceedings of the Sorby Centenary Meeting)*, London, 1964 (Iron & Steel Institute).
13 W. KRUG, J. RIEMITZ and G. SCHULTZ, *The Interference Microscope*, trans. J. H. Dickson; London, 1964 (Hilger and Watts).
14 W. ROSTOKER and J. R. DVORAK, *The Interpretation of Metallographic Structures*, New York, 1965 (Academic).
15 J. NUTTING and R. G. BAKER, *The Microstructure of Metals* (Monograph No. 30), London, 1965 (Inst. of Metals).
16 R. W. CAHN (ed.), *Physical Metallurgy*, chs. 11 and 12, Amsterdam, 1965 (North Holland).
17 R. SMALLMAN and K. H. G. ASHBEE, *Modern Techniques in Metallography*, Oxford, 1966 (Pergamon).
18 D. G. BRANDON, *Modern Techniques in Metallography*, London, 1966 (Butterworth).

ELECTRON MICROSCOPY

19 V. E. COSLETT, *Practical Electron Microscopy*, London, 1951 (Butterworth).
20 *Metallurgical Applications of the Electron Microscope* (Monograph and Report Series No. 8), London, 1951 (Inst. of Metals).
21 *Advances in Electron Metallography*, Vols. 1—6, Philadelphia, 1959—65 (Amer. Soc. Test. Mat.).
22 M. E. HAINE, *The Electron Microscope*, London, 1961 (Spon).
23 G. THOMAS, *Transmission Electron Microscopy of Metals*, New York, 1962 (Wiley).
24 I. S. BRAMMAR and M. A. P. DEWEY, *Specimen Preparation for Electron Microscopy*, Oxford, 1966 (Blackwell Scientific Pub.).

ELECTRON-PROBE MICROANALYSIS

25 P. DUNCUMB, J. V. P. LONG and D. A. MELFORD, *Electron-probe Microanalysis*, London, 1965 (Hilger and Watts).
26 T. D. McKINLEY, K. F. J. HEINRICH and D. B. WITTRY, *The Electron Microprobe*, New York, 1966 (Wiley).

X-RAY MICROSCOPY

27 V. E. COSLETT and W. C. NIXON, *X-ray Microscopy*, Cambridge, 1960 (Cambridge University Press).
28 A. TAYLOR, *X-ray Metallography*, New York, 1961 (Wiley).
29 C. S. BARRETT and E. MASSALSKI, *The Structure of Metals*, New York, 1966 (McGraw-Hill).

FIELD-ION AND FIELD-EMISSION MICROSCOPY

30 D. G. BRANDON, *Modern Techniques in Metallography*, London, 1966 (Butterworth).

Index